FRAMING A HOUSE

ROE OSBORN

FRAMING A HOUSE

ROE OSBORN

The Taunton Press

The Taunton Press
Inspiration for hands-on living®

The Taunton Press, Inc., 63 South Main Street, PO Box 5506, Newtown, CT 06470-5506
e-mail: tp@taunton.com

Editors: Mark Feirer, Peter Chapman

Copy editor: Candace B. Levy

Indexer: Jim Curtis

Jacket/Cover design: Scott Santoro/Worksight

Interior design: Scott Santoro/Worksight

Layout: Emily & Scott Santoro/Worksight

Illustrator: Christopher Mills

Photographer: Roe Osborn except for photos on p. 7: (top right) Chuck Bickford,
Fine Homebuilding © The Taunton Press, Inc., p. 9: (bottom) © Harold Shapiro,
p. 98 (bottom) Andy Engel, *Fine Homebuilding* © The Taunton Press, Inc.

The following name/manufacturer appearing in *Framing a House* is a trademark: Speed®Square

Library of Congress Cataloging-in-Publication Data

Osborn, Roe.
  Framing a house / author, Roe Osborn.
      p. cm.
  Includes index.
  ISBN 978-1-60085-101-8
  1. House framing. I. Title.
  TH2301.O83 2010
  694'.2--dc22
                                    2010018933

Printed in the United States of America
10 9 8 7 6 5 4 3 2 1

Homebuilding is inherently dangerous. From accidents with power tools to falls from ladders, scaffolds, and roofs, builders risk serious injury and even death. We try to promote safe work habits throughout this book, but what is safe for one person under certain circumstances may not be safe for you under different circumstances. So don't try anything you learn about here (or elsewhere) unless you're certain that it is safe for you. If something about an operation doesn't feel right, don't do it. Look for another way. Please keep safety foremost in your mind whenever you're working.

To Judy Megan (1941–2004), who saw me writing about building homes so long ago, and to Bridget Cahill, who saw me through the writing of this book every step of the way.

ACKNOWLEDGMENTS

I owe my appreciation to so many people who took the time over the years to share their knowledge of carpentry and building. Thanks to my dad, who let me watch as he toiled to carve out living space for his five kids while holding down two jobs. Thanks to Rob Turnquist, who had infinite patience while teaching me the carpentry ropes. Thanks to Kevin Ireton, who took a chance hiring me at *Fine Homebuilding* magazine, where I was able to learn from some of the most brilliant and creative minds in the building industry, including Rick Arnold, Mike Guertin, Gary Katz, John and Kerri Spier, and framing guru Larry Haun.

To make this book happen, I sorted through 15 projects before finding a house that fit the aims of the book *and* actually got built in the midst of a brutal economy. Special thanks to Joe Iafrate, the general contractor of the project house, who let me photograph it. Thanks also to the crew on this project: Dan Duberger; Mike Reinville; and the Tripp brothers, Tandan and Troy. Thanks also to Steve Cook, of Cotuit Bay Design, the project designer and engineer, who deciphered the local building code for me.

On the writing end, thanks to Steve Culpepper for suggesting the book, to Peter Chapman for his patience and support in guiding it, and to Mark Feirer, my editor, who slogged through my prose until it made sense.

Thanks to Humphrey and Lily, my two Jack Russells who gave up their dad for the last eight months, though they were always eager to drive to the job site and wait patiently in the car or to hang out in the office while Dad banged on the keyboard until the wee hours.

Finally, and most especially, thanks to my partner and best friend, Bridget Cahill. She was there to dig in the spurs every time I needed a push and always cheered me on . . . and on and on. Always offering a smile, she maintained a steadfast, positive attitude while sacrificing our time together. During the long and arduous process, she supplied me with coffee and made me countless dinners. But most of all, she helped me believe in myself enough to get this book finished.

CONTENTS

INTRODUCTION

I had a good friend who graduated near the top of his class in high school but never bothered with college. All he ever wanted to do was build houses. As soon as he was old enough to swing a hammer for a living, he joined a framing crew. He quickly worked his way through the ranks, absorbing everything he could until he was the foreman. Then he quit. He joined another crew, started at the bottom and again worked his way up to foreman, soaking up all the knowledge of that crew. He quit again and repeated the cycle a half dozen times before finally forming his own company. At that point, he'd seen the methods and mistakes of dozens of craftsmen and had learned the fastest and most effective way to build a house. If you're aspiring to become a professional builder, I'd highly recommend that path.

For me, working as an editor at *Fine Homebuilding* magazine was akin to that process. I tried to tackle every article as if it were my first day on the job. My mission was to soak up the knowledge of the incredibly talented authors it was my pleasure to work with. Sure, I'd done my share of framing in my hammer-swinging days, but I probably learned more behind a camera and in front of a keyboard than I ever did on all the job sites I worked as a carpenter. What I've tried to present in this book is the culmination of that knowledge. I focused on a particular house and a particular crew, and these guys still taught me new things, just as with every other crew.

If you're reading this book, you probably have a gnawing curiosity about how houses are built, with an equally strong desire to gain the knowledge and develop the skills to build your own house at some point. I've lived in half a dozen houses, and in every one I found myself daydreaming about how the house went together and what the job site was like when each house was being built. So I hope this book fuels your curiosity and fires your courage to build your own house. If you do, I can promise that you'll never have a greater sense of pride than knowing that you built the roof over your head. And if working with your hands and building things makes your heart pound with excitement, then building your own house will be an amazingly fun and enjoyable time.

That said, I offer these words of caution.

- Always think through a process before jumping in.
- If there's something you don't understand, find a craftsman who will take the time to explain a process or a procedure. There are many skilled builders who are willing to share their knowledge with people willing to learn.
- If a process doesn't seem safe to you, then it probably isn't. Find a way to do every task safely.
- Always do your prep work before performing a task. That may mean building safe scaffolding or setting up a work table or rerouting hoses and extension cords. Whatever the task, proper preparation will make the job go more smoothly and more safely.
- Wear safety equipment! Professionals are notorious for *not* using proper protection, sometimes out of ignorance and sometimes from a false sense of pride. Don't fall prey to the same weaknesses.

Good luck and build safe!
Roe Osborn

Before You Pick Up That Hammer

A lot goes into building a house. But to me, the most basic and vital part of a house is the frame, the structural skeleton that lends a house its basic look and shape. So before you grab your hammer and saw and dive into that pile of 2×4s, take a moment to become familiar with what a frame does and how it works.

Evolution of the House Frame

In North America, homes made of brick, concrete blocks, stone, and stacked logs can be found in almost every region. In these homes, the material serves as both the structure and the exterior finish. But that's not always the case. For example, a wood-frame house consists of a system of evenly spaced solid-wood pieces covered with wood panels, called *sheathing*, a combination of materials that results in a strong, resilient structure. Here's a brief look at how the wood house frame evolved.

Timber frames

Homes built by the first colonists were often timber frames. This building style, which has enjoyed a renaissance in the last few decades, uses large beams supported by massive posts to create the structure of the house (see the drawing below). Posts and beams connect to each other via complex joinery, often secured by wooden pegs. Timber frames were a natural choice for the first settlers. Large trees were plentiful and had to be cut down to clear land for the house sites and for farmland.

Balloon frames

As land was steadily converted to farming and large timbers became harder to obtain, builders devised a framing system that made good use of smaller, more readily available wood. House frames were fastened together with nails instead of interlocking joinery and pegs, and a carpenter with basic skills could do the work. Instead of heavy, widely spaced timbers, the frame called for lighter pieces of wood spaced more closely together. The first version of this system was called balloon framing and featured long sticks of sawn lumber that extended from the foundation to the roof.

Platform frames

Balloon framing eventually evolved into a stronger method that relied on shorter lengths of lumber. It was called platform framing and is the version we use today. A platform frame consists of wood-framed walls capped with a floor system, or platform. As a house increases in height, the platform serves as the base for

Platform-framed houses use the same lightweight wall framing as balloon frames, but the lumber is only a single story tall. The wall members are capped with a platform, or floor, to create a sturdy boxlike structure.

History of House Frames

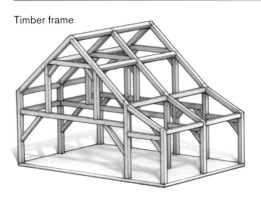

Timber frame

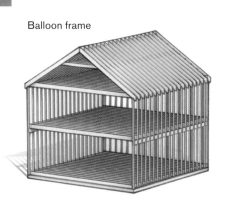

Balloon frame

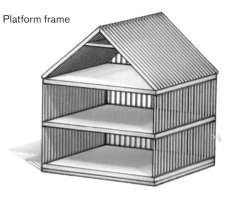

Platform frame

the next set of walls. The single-story lengths of lumber used in platform framing are less expensive, easier to produce, and more widely available than the lumber needed for balloon framing. The frame can be built very quickly, and when sheathed, it takes on the structural qualities of a stiff, solid box. Platform framing is the framing method described in this book.

The Building Code: Framing Ground Rules

To protect the inhabitants of our houses as well as the people who build them, houses are built according to a set of rules called *building codes*. These codes spell out the *minimum* requirements for most aspects of building a house. Code stipulates the size and grades of the lumber and fasteners that go into the frame as well as how the frame should be assembled to handle all the forces that affect it. Building according to code is your assurance that the house will be strong and safe.

Transferring Loads

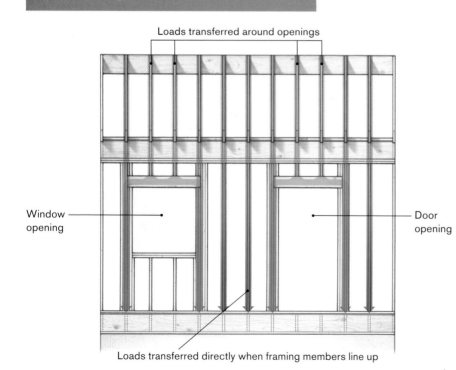

Loads transferred around openings

Window opening

Door opening

Loads transferred directly when framing members line up

Building codes once varied considerably from region to region, but efforts to standardize building practices have led to the development of the *International Residential Code* (IRC), which was first published in 2000. The IRC is a model code, a template that states can adopt in whole or in part. But even if a state adopts the IRC, local jurisdictions generally amend it to fit their specific conditions (see "Codes for Special Locations" on the facing page). A town in a particularly hazardous fire zone, for example, might decide to make local codes more stringent than the IRC regarding acceptable roofing products.

The IRC has now been wholly or partially adopted at the state or local level by all 50 states and the District of Columbia, but you must always follow the specific regulations in your area. *Always make sure that your plans reflect any local requirements that may differ from the general building code.* Be sure to check with your local building department for information. If your local code differs from anything shown in this book, build according to your area's requirements.

The Importance of Uniform Spacing

The frame gives the house its shape and provides structural support, but there's much more to it than just holding the roof up. A modern platform frame is a carefully designed system that depends on the precise and consistent location of each piece. The result is a predictably strong, impressively versatile structure. When framing, never forget that your work creates the base for other crews. An accurately built frame makes life easier for those who will complete the house, from the insulation installers and the drywall guys to the finish carpenters.

Structural support

A house frame transfers the weight of a house to the foundation. Starting at the roof, the loads follow lines of support through the rafters, joists, and wall studs. Loads are transferred efficiently when the framing

The stud and joist bays of a platform frame provide mounting locations for the house's vital systems, such as plumbing, electrical, and heating.

Fiberglass insulation is the most common type of insulation used in houses. Uniform spacing of wall studs makes it possible to install the batts efficiently.

elements line up with each other. But a house is not a sealed box, and those lines of support are often interrupted by openings for doors and windows. The frame must transfer the loads around those holes. If the framing is not uniform, the loads will be distributed in unpredictable ways, which could cause the structure to fail.

Support for systems and materials

The house frame also supports the building's plumbing, electrical, heating/cooling, and communications systems. The spaces between framing members, called *bays*, allow the systems to be hidden from view, so pipes and wires won't become a part of your house's décor. Framing that's not uniform makes installing these systems more difficult and more time-consuming. Consider insulation, for example. Say your framing layout is a little off, and a few stud bays are 17 in. wide instead of 14½ in. (the distance between 2× framing lumber at 16 in. on center, and the norm for most framing projects). In this case, standard insulation batts won't fit, which means somebody has to figure out how to solve the problem—and that takes time and will probably waste material or result in a less energy efficient wall.

The frame of a house provides a solid base for attaching finish materials, such as corner boards and soffits, and weatherization materials, such as roof shingles.

Codes for Special Locations

The house described in this book was built in New England, in a high-wind area near the ocean. Local codes require all new construction to conform to a 110-mph-wind designation. This means that houses built in this area must be able to withstand sustained winds of that strength. Throughout the book, I highlight examples of how this designation changed the framing requirements for the house.

The frame of a house also provides a solid place to attach finish materials. Outside, the roofing, siding, and exterior trim are secured to the frame. Inside, the framing affords solid support for drywall, cabinetry, and trim.

Before the Frame

Plenty of work has to occur on the site before the frame can be built, including providing access and temporary power. The quality and accuracy of the site work and the foundation greatly affect the framing. If those go wrong, you'll know it as soon as framing starts.

A pretty site is a pretty sight. This foundation (not the project house) has been back-filled, making it possible to stack the framing materials neatly. No excavation or formwork debris remains.

Site work

Good excavators do their work in the least invasive way to protect trees and other natural features, so choose a good one (not necessarily the cheapest one). A site that's been prepared by a good excavator is much safer to work around. Before you begin building, the excavator should remove debris, including downed trees and brush, from the lot and should pile topsoil cleared from the lot neatly and safely away from the foundation to give you plenty of clear, flat area to work and to stack framing materials.

Whether to backfill the foundation before framing the first-floor deck is a matter of considerable debate among builders. On one hand, filling the trenches around the foundation provides easier, more convenient, and safer access for crew members. But many builders opt to backfill after framing the first-floor deck because backfilling can cause excessive pressure on new concrete, especially if the foundation walls are long and are not braced on the inside. Floor

A back-filled foundation lets the crew work on the first floor while standing safely on the soil (below, left). However, some contractors hold off on backfilling the foundation to give the concrete extra time to cure completely, which can make the initial stages of the project somewhat challenging (below, right).

framing stabilizes foundation walls and greatly reduces any chance of failure from the pressure of backfilling (see chapter 3).

Foundations

A foundation anchors a house to the earth and supports its weight. The type of foundation you choose varies, depending on your geographic location, the topography of your building site, and the type of soil on the site. Building a foundation is an art form in itself and has been the subject of many books. Here's a brief overview of some common foundation types and my take on the pros and cons of each one.

Full-basement foundations

Full-basement foundations are good choices for areas of the country where the ground freezes in winter. A full basement features either poured concrete or concrete block walls that are typically 8 ft. to 9 ft. high and create a basement space with the same footprint as the house.

The best part about a full basement is the extra area it provides. It can be used for storage as well as for utilities such as the boiler and water heater. A full basement can also be finished to create living and working space or even an extra bedroom, if there is sufficient egress (a way out in case of emergency). On the downside, full basements add to the cost of the house because of the amount of excavation, forming, and material required. Full basements can also be dark and prone to moisture problems, making them uninviting locations for extra storage.

Crawl-space foundations

If you live in an area where the ground is subject to freezing but don't want or need a full basement, you might opt for a crawl-space foundation. The concrete or block walls that form this foundation are called frost walls or stem walls. The excavation needed to prepare for them is easier, the foundation forms take less labor, and there's less concrete to pour or block to lay, which can add up to a substantial savings compared to a full basement.

Furthermore, the area between the foundation walls is often bare soil, which saves the cost of pouring a basement slab. However, bare soil should always be sealed by some other method to discourage moist conditions that can lead to mold in the house and rot in its framing. Another disadvantage of a crawl-space foundation, whether or not it has a dirt floor, is the inconvenience of having to work on wiring and plumbing in a claustrophobic space, usually while you're on your back trying to balance a flashlight on your chest.

Full-basement foundations such as this one require extra material but create a large amount of useful space below the main living levels of a house.

Crawl-space foundations save material but leave a minimal amount of area between the ground and the floor framing. Support posts rest on piers or footings. Bare soil should be covered with a vapor barrier (being installed here).

Slab foundations Concrete-slab foundations are common in mild climates where the frost line is relatively shallow. They can be installed quickly and are often supported by short stem walls that are usually formed and poured separately. In warm areas, such as the desert Southwest, excavation and installation are even easier. The surface soil is scraped away, shallow trenches are excavated for footings, and the slab and stem wall are poured as a monolithic unit with built-up edges to support and distribute the weight of the house.

The biggest advantage of a slab foundation is that the house walls are connected directly to the slab, eliminating all the material and labor costs associated with floor framing. But the biggest drawback is the extra work needed to prep and form the slab. Chases for wires and other systems have to be installed and braced in exactly the right places before the concrete truck arrives. Modifications after the slab is poured are difficult and often result in unsightly and uneven patches. Another drawback is that a concrete floor isn't very comfortable under foot or under child.

Pier-and-grade-beam foundations A pier-and-grade-beam foundation consists of deep piers or columns that support a series of horizontal beams below the house. The floor of the house then attaches to the beams so that the house can nestle into a hillside or problematic site. This type of foundation lets you make use of a challenging building lot, such as a steep slope, where any of the more standard types of foundations would not be practical. But besides the rigorous engineering required for a pier-and-grade-beam foundation, the biggest disadvantage is working on a site where many of the standard building procedures simply do not work.

Components of a House Frame

Before you frame a house, you need to know the parts of the frame and how they work. The following discussion roughly reflects the order in which the parts are built in the framing process and are discussed in the book.

Concrete slab foundations are a common choice in mild climates. The foundation and the first floor are completed in a single monolithic pour.

Parts of the Frame

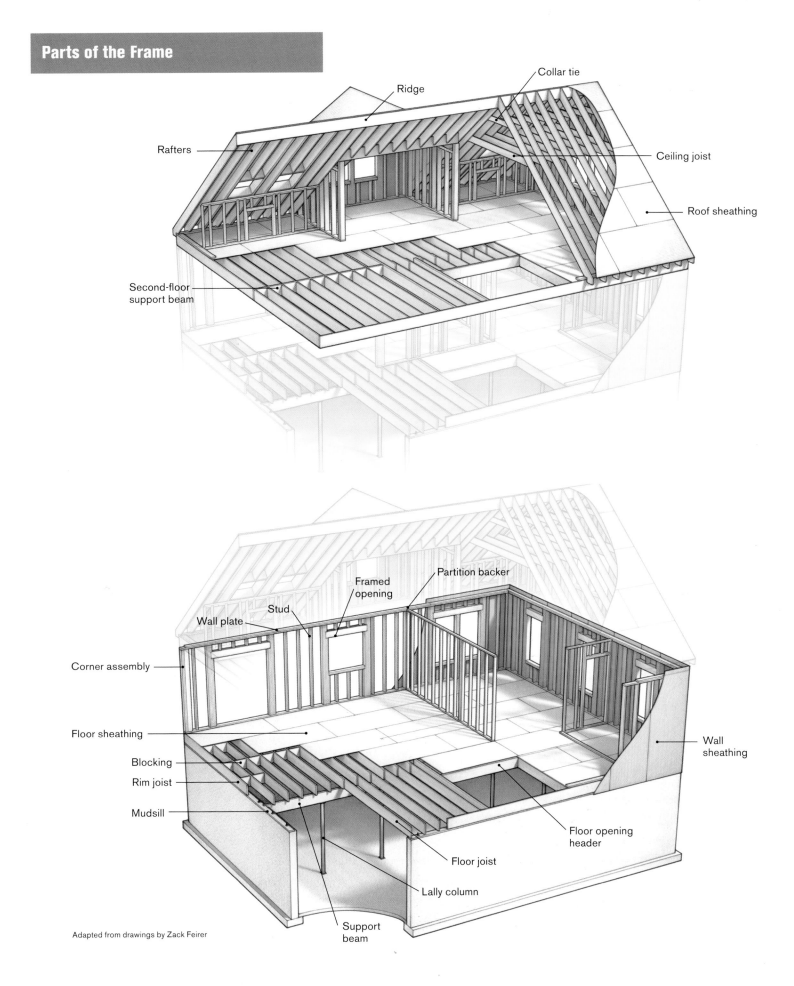

Ridge

Collar tie

Rafters

Ceiling joist

Roof sheathing

Second-floor support beam

Framed opening

Partition backer

Stud

Wall plate

Corner assembly

Floor sheathing

Blocking

Rim joist

Mudsill

Wall sheathing

Floor opening header

Floor joist

Lally column

Support beam

Adapted from drawings by Zack Feirer

Mudsills

Pressure-treated lengths of 2×6 lumber, which form the first layer of the mudsill, are solidly attached to the foundation with bolts that have been embedded in the foundation walls. They provide a bearing surface that

Mudsill

prevents contact between the untreated floor joists and the concrete foundation. A second layer of standard lumber is sometimes applied over the pressure-treated boards. That layer can be used to help level the mudsill, and the light-colored wood makes the layout marks easy to read. However, most mudsills are built with a single layer of pressure-treated lumber.

Support beam

Also called a center beam, the support beam bears the first-floor joists if they are not long enough to span the entire distance between foundation walls. The ends of the support beam bear on the foundation walls; intermediate support below the beam is usually provided by Lally columns (hollow-metal columns filled with concrete). The top of the beam should be on the same plane as the top surface of the mudsills.

Floor joists

Floor joists are horizontal lengths of lumber placed on edge to span the area between the mudsills and the support beam. The joists must be nailed to the support beam, the rim joists, and the mudsills.

Floor joist

Rim joists

Rim joists are the same material and dimension as the floor joists but rest entirely on the outer edge of the mudsills. They must be nailed both to the mudsills and to the ends of the floor joists. They close off the ends of the joist bays and help hold the joists in a vertical position.

Support beam

Rim joist

Floor header

Some of the floor joists must be cut short to create openings for features such as stairs and chimneys. A floor header supports the ends of the interrupted joists. Headers are often made of two or more lengths of floor-joist stock sandwiched together but can also be made of manufactured material such as laminated veneer lumber (LVL), as in this house. To carry the extra load placed on them by the headers, the floor joists on either side of the opening must be doubled or tripled.

Floor header

Blocking

Building codes often require that short pieces of lumber, called blocking, be nailed between joists and rafters to reinforce or stiffen the floor or roof system. When blocks are installed in a row at the midspan of the joists and run the entire length of the building, they are called bridging. Blocking can also be installed between wall studs to make it easier to support such things as handrails, towel bars, and cabinets.

Floor sheathing

Floor sheathing is a layer of plywood or oriented-strand board (OSB) that is nailed and glued to the top edge of the floor joists. Floor sheathing stabilizes the joists and provides a solid, flat base for wall framing and finish flooring.

Blocking

Floor sheathing

Exterior wall

Wall plates   Stud   Partition backer

Exterior walls

Exterior walls are framed with 2× material covered with sheathing and siding on the surface exposed to the weather. These walls are usually insulated.

Interior (partition) walls

Interior walls subdivide the inside space of the house, separating rooms and creating closets. These walls are not usually insulated.

Wall plates

Wall plates are the horizontal members at the top and bottom of each wall; they tie the ends of the studs together. The bottom plate is nailed through the floor sheathing and into the joists below. The top plate supports floor joists or ceiling joists. It is common practice to double the top plate in order to tie adjacent walls to each other and provide extra support for second-story floor joists.

Studs

The 2×4 or 2×6 vertical members of both exterior and interior walls are known as studs. They extend between the top and the bottom wall plates. Unlike other framing members, studs can be precut at the mill to one of several standard lengths and are generally installed without further cutting.

Corner assembly

A corner assembly consists of two or three studs nailed together in an L shape to form a rigid corner where framed walls meet. The assembly also provides a stiff, solid nailing surface for attaching interior wall finishes.

Partition backer

A partition backer is an assembly of three studs nailed together into a U shape. When built into a wall, the backer creates a stiff column to which an intersecting

Corner assembly

wall can be attached. The legs of the U stiffen the backer stud and provide a nailing surface for interior wall finishes.

Framed opening

Openings in walls that accommodate windows or doors consist of several pieces of framing. The full-length studs on the sides of every opening in the wall are called king studs. The somewhat shorter studs that support the ends of a header are called trimmer studs or jack studs. They are nailed to king studs or to other jack studs. A header supports framing above and distributes loads to the jack studs on each side of the opening. The header fits on top of the jacks and between the king studs. A stool (also called a sill or saddle) is a horizontal member at the bottom of a window opening that fits between the jacks. Short framing members that fit under the stool or above the header are called cripple studs. Local building codes may require extra framing members around openings to stiffen them against high winds, as shown in the left photo below.

Wall sheathing

The layer of plywood or OSB that skins the exterior of the wall is called wall sheathing. It adds lateral strength to the wall system by tying the framing members together. It also serves as a nailing base for siding.

Framed opening

Header

King stud

Jack stud

Cripple stud

Wall sheathing

Ceiling joist

Ceiling joists

Ceiling joists are similar to floor joists, except that they do not support conditioned living space above. Their primary purpose is to provide a nailing surface for ceiling materials such as drywall. Because they don't carry floor loads, ceiling joists do not need to be as wide as the floor joists, although they usually are. Ceiling joists can go between the rafters, or finish the top of a platform with rim joists around the perimeter.

Rafters and Ridge

Angled framing members that create the shape of the roof are called rafters. They support the weight of the roof sheathing and the roofing material and channel the loads to the exterior walls. The board that joins the ends of the rafters at the peak is called the ridge. It is often made of 2× lumber but can be made of other materials. A *structural ridge* is supported by posts and helps carry the weight of the roof. A *nonstructural ridge* does not carry roof weight. Both types of ridges provide a bearing surface for the ends of the rafters and tie them together. Rafters also provide a nailing surface for the roof sheathing.

Roof sheathing

The plywood or OSB skin that covers the rafters and joins them together is known as the roof sheathing.

Collar ties

Collar ties are horizontal members that join opposing rafters together. They stiffen the roof and can serve as ceiling joists when the attic is used as living space.

Rafters and ridge

Roof sheathing

Collar tie

Frame Engineering

You can spend years in school to become a structural engineer, but you don't need that kind of training to build a house. Still, a general working knowledge of the forces at work in a house frame and on a frame can be very helpful. Awareness of those forces and how the frame resists them informs your choice of techniques and makes strict adherence to the building code more natural and intuitive, especially in jurisdictions where enforcement of regulations might be more relaxed.

Loads and Span

The way a frame reacts to the various forces applied to it depends a lot on factors such as how heavily it is loaded and the distance the horizontal framing elements must span.

A *load* is the weight carried by the framing members. Loads are separated into two categories: dead loads and live loads. Dead loads are the fixed weight of the structure, including the materials and any fixtures or equipment that are a permanent part of the building. Live loads, on the other hand, consist of any changeable weight that the structure is subjected to, such as people and furniture. Both types of loads must be taken into consideration when determining the proper size and spacing of framing members.

The distance that a framing member can extend between supports is called the *span*. The distance a particular piece of lumber can span is affected by various factors, including the magnitude and location of the loads it must bear. It is also affected by the species of wood. A 2×6 made of oak can span a greater distance than one made of pine because the material itself is stronger. Span tables are available both online and at most municipal building offices. Although the tables apply to general floor framing, in some areas of the country, plans must be reviewed by an engineer.

Forces at work in a frame

A house looks as though it were simply resting peacefully as it sits on its lot. But beneath the clapboards and shingles, the frame is hard at work resisting forces. Most of these forces have to do with gravity, and the frame must be engineered in such a way that it handles them easily and efficiently. It also has to be ready for forces that come along only once in a while, such as heavy wind or seismic activity.

Deflection Deflection is the amount a framing member or a framing system, such as a floor or roof, moves downward when weight is added. Deflection is affected by the spacing of the lumber. For example, standard framing members spaced 16 in. on center (o.c.) might deflect less than wider framing members spaced at 24 in. o.c. Excessive deflection can result in cracked drywall or a cracked tile floor.

Tension and compression The two forces of tension and compression are the opposite of each other. Tension is a pulling or stretching force, whereas compression is a pushing or squeezing force. For example, in a roof with a nonstructural ridge, the weight of the rafters on the plates acts to spread the walls apart. However, the joists below each rafter keep that from happening by pulling the walls together. The joists are then said to be in tension. At the same time, the weight of the floor system pushing down on the studs tends to squeeze them, putting them in compression.

Shear Shear is a force applied perpendicular or opposite to the normal resistance strength of a structure or material. That force could be wind pushing on the side of a building, but it could also be a lean-to roof pushing against studs in a wall. Another example of shear is the downward force on the end of a cantilevered joist. Shear forces can be some of the most subtle yet most destructive forces.

Structural Forces

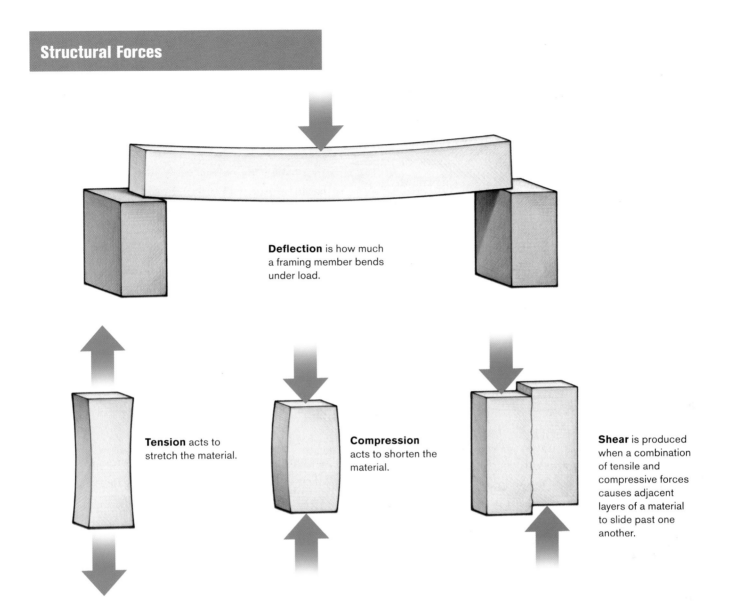

Deflection is how much a framing member bends under load.

Tension acts to stretch the material.

Compression acts to shorten the material.

Shear is produced when a combination of tensile and compressive forces causes adjacent layers of a material to slide past one another.

Nailing schedules

Contrary to what the name implies, a nailing schedule has nothing to do with *when* you drive nails. Instead it has to do with how many nails and what size nails you must use for a given application. Building codes include general nailing schedules for most framing connections. But if you're building in an area with special code requirements, such as for high winds, a nailing schedule should be included as part of the house plans. The minimum requirements of a nailing schedule must always be followed exactly to ensure the strength of the connections. These details are especially important in certain applications, such as nailing sheathing to the wall framing (see chapter 4).

The size and spacing of the nails must be followed precisely or the house will not meet local code requirements. Building inspectors will flag this error in a heartbeat.

Framing Materials

No preamble to the process of framing would be complete without a discussion of framing materials. The house frame has evolved and so too have the materials that go into building it.

Dimensional lumber is solid wood framing material cut directly from trees. The process of surfacing each piece reduces the nominal dimensions by about ½ in. in each direction. For example, a 2×4 is actually 1½ in. by 3½ in.

Lumber Grades

As lumber is manufactured, it goes through a rigorous system of inspection and grading. Knots that are too big, cracks or checks in the grain, and many other defects can cause a piece of lumber to be rejected. After the lumber has been inspected, it is given a grade stamp. That stamp indicates the species of wood as well as how the lumber was prepared and stored—for example, green, stack-dried, or kiln-dried (each of which can cause the moisture content to vary). Green lumber tends to be the most unstable and the most liable to warp or twist. Most framing lumber will never be seen, so appearance is not an issue. However, with finish lumber or exposed framing lumber, an appearance grade is also included. The best lumber is Clear (it has no flaws), followed by Select (a few small, tight knots), #2 (larger tight, even knots), and #3 (a variety of knots and minor flaws).

Dimensional lumber

For most of the last century, the industry standard for framing material has been dimensional lumber (solid wood milled to uniform dimensions in standard lengths). It's easy to cut and readily available, and its long track record means that framers know it well. Of course, no material is perfect, and dimensional lumber does have its faults. Wood tends to expand or shrink somewhat as it absorbs or sheds moisture. These changes tend to reflect the seasons in many climates, with wood drying out and shrinking in the winter and then expanding with the moist, humid conditions of summer. That's why cracks in walls or ceilings sometimes appear and then mysteriously disappear, and why squeaks sometimes develop in floor systems. Excessive moisture content can eventually cause wood to decay but it's not the material's only enemy. Many people, particularly those in the mild climates, are uncomfortably familiar with a host of wood-destroying insects. Good thing wood readily absorbs chemicals that help it to resist decay and ward off insects.

Common nominal dimensions for framing lumber are 2×3, 2×4, 2×6, 2×8, 2×10, and 2×12. These lumber sizes are usually available in lengths of 8 ft., 10 ft., 12 ft., 14 ft., and 16 ft., although all lengths may not be available for all widths, and longer lengths may be available for some.

In the course of framing a house you'll also find 1× lumber used for trim, 4× lumber for posts, and 6× stock for headers and beams. These are the thicknesses of lumber that the sawmills mass-produce in various widths for residential construction, and that's mostly what we used for the house described in this book.

Engineered lumber

Dimensional lumber is solid wood cut directly from a tree into specific sizes. Engineered lumber, on the other hand, starts as solid wood but is further processed into fibers, strands, chips, or veneers that can be glued together in various forms to create a wood-based material that is stronger and more stable than the original wood. For example, layers of veneer glued together form sheets of plywood or laminated-veneer lumber (LVL). Strands of wood mixed with glue are

Understanding Lumber Dimensions

The most common dimensional, or framing, lumber is known as 2× or 2-by stock, the stuff of studs and joists and rafters. This term refers to its thickness and is a nominal dimension, not an actual measurement. By the time the lumber has been milled so that it has smooth surfaces and consistent dimensions, its actual thickness is typically 1½ in. That's why the nominal and actual widths of lumber differ. For example, a 2×4 (the nominal dimension) is really 1½ in. thick and 3½ in. wide (the actual dimensions).

known as oriented-strand board (OSB) panels or lumber. Wood chips and glue become particleboard. The I-joist is an increasingly common alternative to dimensional lumber. It is a strong, lightweight framing member, usually consisting of an OSB web and solid wood or LVL flanges.

I-joists offer a lightweight alternative to lumber. Consisting of a vertical web capped with flanges, I-joists are manufactured in many sizes and lengths.

LVL beams, engineered lumber products, are much stronger than dimensional lumber.

Because engineered lumber has been precisely fabricated in a factory, it has none of the knots, splits, or other natural flaws common to dimension lumber. Best of all, it's usually dead straight, very predictable, and less prone to shrinkage and expansion. Because these are engineered products, however, you must follow strict guidelines for working with them or you may compromise their structural integrity. Always check the manufacturer's guidelines when working with engineered lumber.

The house described in this book was built with dimension lumber, but the plans called for LVLs to be installed in strategic places to add strength to the floor framing in areas of extra stress. LVLs have become common on residential job sites. They are many times stronger than similar-size dimensional lumber and are available in much longer lengths. However, an LVL is quite heavy and may require special fasteners.

Sheathing Before World War II most houses were sheathed with some form of solid wood, usually 1× boards that were nailed in place either perpendicularly or diagonally across the framing. Board sheathing is long gone. Plywood was the first to replace it, followed by OSB. Both are engineered wood products, and their large size speeds the work of sheathing a house. Plywood was long the industry standard, and when OSB first became available it gained a reputation for deteriorating quickly when exposed to moisture. However, OSB manufactured today is every bit as strong as plywood and stands up to moisture exposure just as well. The two are usually comparable in price, but OSB is heavier and therefore a bit harder to handle than plywood. The floor sheathing chosen for the project house is OSB; the walls and roof were sheathed with plywood.

Nails

Most of the connections in a house frame are made with a relatively small range of nail types and sizes, but it's critically important to use the right ones. In fact, building codes specify the size, type, and number of nails that must be used for each connection in the frame.

Nail shanks come in a variety of lengths and thicknesses that are designated by a penny scale. The letter *d* following the number is the abbreviation for "penny." The longer and heavier the shaft of the nail, the higher the penny number.

When it comes to nailing, however, more is not necessarily better. Excessive nailing can split the wood and actually weaken a connection, so get familiar with the building code requirements for nailing.

Nail materials and finishes Most nails are made of steel but may have various finishes. Steel nails with no coating or finish are called bright nails. For most framing connections that aren't exposed to the outside, bright nails are sufficient, and they're usually less expensive than others. In most areas, coated framing nails are available. These nails have a vinyl or resin coating that makes them easier to drive. The coating melts to provide lubrication as the nail is driven, and then solidifies to increase the nail's holding power. Most pneumatically driven nails are coated.

Sheathing consists of panels or sheets that cover either the studs, the rafters, or the floor joists. It is usually made of plywood or OSB.

Standard Nail Sizes

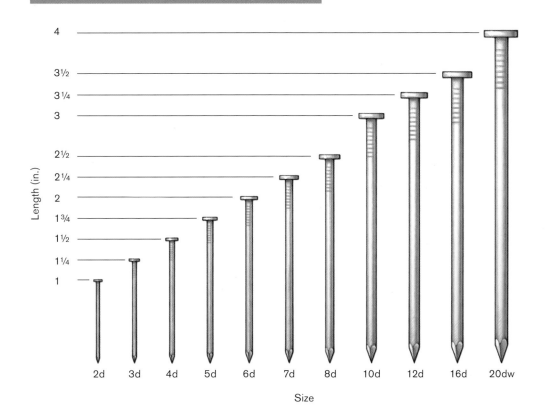

Pneumatic Nailers and Code

When pneumatic framing nailers first came out, manufacturers had to figure out how to fit the maximum number of common nails in the tools. The solution was the D-head (clipped head) nail. The D shape allowed the nails to fit tightly together, shank against shank. If you use D-head nails or any other special nails for pneumatic tools, make sure they are approved by your local code. Some jurisdictions may restrict their use.

Choosing Galvanized Fasteners

The amount of zinc on a galvanized nail can vary, as can the method used to apply it. The combination of these factors determines the corrosion resistance of the nail. Electroplated nails have a thin, but uniform, coating of zinc. Hot-galvanized nails are tumbled in a drum with small bits of zinc that melt on the nails as the drum is heated. Hot-dipped nails are dipped in molten zinc. Electroplated nails are the least durable of the three; hot-dipped nails are the most durable. Make sure the galvanized nails you choose are labeled specifically for your intended use. Expect to pay a premium for the most corrosion-resistant galvanized nails.

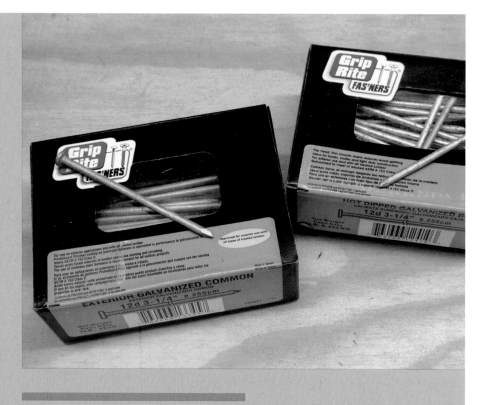

Galvanized nails aren't all created equal. Be sure to check the label for your intended use.

Steel nails intended for use in exterior applications often have a coating of zinc that inhibits rust, applied in a process called *galvanizing*. The zinc also increases the holding power of a nail. However, not all galvanized nails are equal in their corrosion resistance (see "Choosing Galvanized Fasteners" on the facing page).

Stainless-steel nails are available but they're much more expensive than standard steel nails or even the best galvanized steel nails. But if you're building where corrosion problems are severe, such as close to the ocean, consider an investment in stainless-steel as your best chance to eliminate corrosion.

Basic types of nails

There are five basic types of nails:

COMMON NAILS Nails with large heads and thick shafts are called common nails. The large head on a common nail gives it superior holding ability and makes it easier to drive with a hammer. Most framing connections are made with 12d or 16d common nails.

BOX NAILS Thin-shank nails with large heads are called box nails. In many parts of the country, 12d and 16d box nails called sinkers are used for framing connections. Compared to common nails, the thinner nail shaft makes them easier to drive by hand and less likely to split the wood.

FINISH NAILS AND CASING NAILS Nails with very small heads are called finish nails. The small head of a finish nail makes it less conspicuous where the hold is less important than the looks. Finish nails are easy to set below the surface of the wood. The hole can then be filled to make the fastener completely invisible. A casing nail is like a heavy-duty finish nail and has a somewhat wider, conical-shaped head. Its extra heft makes it a good choice for applying exterior trim.

SHEATHING NAILS Floor sheathing and wall sheathing require particularly tenacious connections. Extra holding power does not come from making the nails

larger or longer. Instead, rings are formed into the nail shank to give the nails extra withdrawal resistance. Ring-shank nails for sheathing are usually 6d or 8d.

JOIST HANGER NAILS Joist hanger nails are used for fastening metal connectors. This type of nail has an unusually stout shank and relatively short length to resist shear forces. Metal connector nails also have thicker heads that are less likely to pop off the nail shank.

Metal connectors

To fasten a joist or a rafter to a header, nails alone may not be strong enough; your building code may require that you use a metal connector known as a joist hanger. These connectors have become ubiquitous in part because they significantly strengthen the connections. Joist hangers and other connectors, identified by size, type, material, and finish, will be specified on the plans wherever they are required.

How Many Nails Will You Need?

Nails are cheapest when you purchase large quantities. If you're hand-nailing the frame of a small house, start with one 50-lb. box each of 12d and 16d commons and buy more as needed. Buy an equivalent quantity if you're using a pneumatic nailer. You'll also need a couple of different lengths of finish and galvanized casing nails, so start with smaller quantities of those (boxes of 1,000) and purchase more as needed. Stainless-steel nails are the most expensive, so buy them in smaller quantities and treat them like gold.

Metal connectors such as joist hangers reinforce the unsupported joints between framing members.

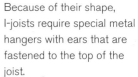
Because of their shape, I-joists require special metal hangers with ears that are fastened to the top of the joist.

Joist hangers come in many sizes to fit various dimensions of lumber, including doubled and tripled 2×s. Some metal connectors solve special problems—for example, one type has a flexible bottom flange used to connect shed-roof rafters to walls. I-joists have dedicated hangers with ears, or top flanges, that go over the top of the joist.

Metal connectors have been known to deteriorate when exposed to the chemicals found in pressure-treated wood. Use galvanized and stainless-steel joist hangers, which resist corrosion, but make sure the product is specifically rated for use with pressure-treated wood.

In addition to joist hangers, many local codes now require metal connectors in areas such as between the rafters and the wall plates to strengthen the structure against uplift forces caused by high wind. They come in many different configurations to fit a variety of framing situations.

In regions of the country that are often pummeled by high winds, metal connectors are required by building codes to create a strong bond between the rafters and the wall plates.

2

Basic Framing Tools and Tool Techniques

When platform framing became the home-building method of choice, it simplified the construction process and sparked the development of many new construction techniques. Meanwhile, builders had to keep up with advances in technology as well as learn to work with new types of tools that began to arrive on job sites. This chapter familiarizes you with those techniques and tools and gives you tips for making the processes move quickly and efficiently.

Measuring and Layout

In the days of timber framing, layout meant determining how all the pieces ought to go together and then drawing complex joinery directly on the timbers so craftsmen would know where to make their cuts. With platform framing, layout means locating the position of and ensuring consistent spacing of every component, including joists, studs, and rafters.

Making your mark

You could be forgiven for overlooking the humble carpenter's pencil when making a list of important tools, but few houses get built without one. Standard round or hexagonal pencils break easily on a job site. Moreover, they need frequent sharpening and can't be set down without rolling away. That's why the pencil of choice is a carpenter's pencil, a flat stick of ¼-in.-thick cedar surrounding a hard, flat lead. The pencil can be sharpened quickly with a utility knife, sort of like you'd whittle a flat twig.

Sharpen a carpenter's pencil with your utility knife. Grip the pencil firmly and push on the back of a utility knife with your thumb.

An open-reel tape (top) is used to measure long walls and check a foundation for square. A 25-ft. tape (bottom) is the tool for other measuring tasks during house framing.

Using and caring for a tape measure

Tape measures come in many sizes, but you'll need two for framing layout: a 100-ft. open-reel tape for long measurements such as checking foundations and a 25-ft. tape for everything else.

A 25-ft. tape is absolutely indispensable and will cover 90 percent of the measurements you'll make while framing. Some guys use 30-ft. tapes, which are heavier and bulkier, but most people have trouble pulling out the last couple of feet from them anyway.

I once worked with a guy who thought the end of his tape measure was broken because the hook was loose and slid back and forth. Actually, this play is a design feature built into every tape. When you hook the end of your tape on a board, the measurement you read starts at the *inside* surface of the hook (see the top photo on p. 30). When you butt the tape against something solid for the measurement, you're taking the measurement from the *outside* of the hook (see the bottom photo on p. 30). In other words, in order for the tape to read accurately, the hook must be able to slide a distance equal to its thickness.

ESSENTIAL TECHNIQUE

Protecting Your Tape

Tape measures are self-retracting, meaning that a spring inside reels the blade back into the housing after it's extended. But the quickest way to damage or break the tool is to retract the tape at full speed and let the hook slam into the housing. To extend the life of your tape, always slow the retraction rate, either by guiding the tape with your fingers or by feathering the locking mechanism.

In any case, your tape measure should be a heavy-duty model with at least a 1-in.-wide blade. Heavy-duty tapes usually have a metal or a high-impact plastic housing, and they have stronger, longer lasting springs to rewind the tape. Many of them also have reinforcement on the first few inches where the tape is liable to wear out first. A wide blade lets you extend the tape straight out for 10 ft. or more without it buckling, a seemingly insignificant feature that you'll soon find invaluable.

Longer 50-ft. or 100-ft. tapes are necessary when measuring long walls or taking long diagonal measurements. These tapes have an open reel (the tape roll is exposed), or they fit into an enclosed case. All have a hand crank to reel the tape back in—the spring needed to retract a 100-ft. tape would make the tool heavy and cumbersome. The tape itself is often made of fiberglass instead of steel to save weight.

The slight play at the end of the measuring tape means you'll get the same number whether you hook the tape on the end of a board (top) or butt the tape into a flat surface at the end of the board (bottom).

Measuring tapes with wide blades can be extended over 10 ft. before the tape threatens to buckle.

Tape Measures and Moisture

In a perfect world all houses would be framed in sunny weather, but you'll often find yourself working in the rain or snow or heel-deep in a muddy quagmire. If a tape retracts with mud or water on the blade, the tool's life will diminish quickly and dramatically. Moisture causes the retraction mechanism to rust, and muddy grit eventually scours the tape until you can't read the numbers. In wet weather, wipe the blade as it retracts. With proper care, a tape can endure the rigors of several framing projects before it should be retired.

Measuring tapes have marks spaced every 16 in. to help with framing layout. Foot marks are also emphasized.

Using a tape for layout Pull out a tape and you'll notice that every foot mark is emphasized, but also that the numbers are given a special mark every 16 in. These marks guide you when laying out studs, joists, or rafters spaced 16-in. o.c. But here's the catch: If you just made a mark every 16 in. starting from the end of the plate, the end of the sheathing would fall on the edge of the board instead of in the center, and extra nailers would be needed to make sure the sheathing had proper attachment.

There are two ways to get around this problem. The first method is to mark back half the width of the stud (¾ in.) at every position. So instead of marking 16 in., 32 in., 48 in., and so on, you'd mark 15¼ in., 31¼ in., and 47¼ in. The second method begins with measuring out 15¼ in. Next partially drive a small nail at that mark. Then just hook your blade on the head of the nail—that's what the little cutout in the hook is for. Now it's easy to mark your layout according to the tape's on-center markings.

Most tapes also have marks for 19.2-in. spacing, although in all my years on job sites, I've never seen anyone use them. The marks are essentially a compromise between a 16-in. layout and a 24-in. layout. In a distance of 8 ft., a 16-in. layout creates six spaces with seven framing members, and a 24-in. layout creates four spaces with five framing members. If you lay out 8 ft. at 19.2 in. o.c., you'll have five spaces with six framing members. This method eliminates one framing member compared to the 16-in. layout, and it's stronger than the 24-in. layout.

Using a triangular square

Once the positions of the framing members are marked, a triangular square (the Speed® Square is one brand) is used to square a line—draw a line 90 degrees from the edge—across the board at each mark. A triangular square has a lip along one edge. When the lip sits against the edge of a board, the edge of the square runs perpendicularly across the width. The lip also makes it easy to slide the square along a board to draw square lines repeatedly.

Unlike a traditional L-shaped framing square, a standard (7 in. on a side) triangular square is small enough to fit in most tool belt pouches. When framing rafters, a larger triangular square (12 in. on a side) works just as well as a framing square, and it won't go out of alignment if you drop it. If you want to make a

perfect 90-degree cut across a board, use the triangular square to guide the saw. It makes a pretty good tool for scraping ice off of lumber too.

I'll discuss specific techniques for using a triangular square as they come up during the course of this book. The tool is particularly useful for streamlining the layout of rafters (see chapter 8).

Hammers (the Original Cordless Nailer)

Like many carpenters, my very first day on the job was spent running baseboard in closets. I was visibly frustrated with not being able to drive nails straight,

and finally Rob, the lead carpenter, came over and handed me his hammer. They say a poor carpenter blames his tools, but I could not believe how easily the nails went in after switching hammers. That night I bought my own professional hammer and have used one ever since.

Even with the best pneumatic tools, you'll always need a good hammer when framing a house. But *good* is partly a matter of personal preference. Long-time framer Larry Haun swears by his wood-handled hammer, but it feels unbalanced and alien in my hand. He'd probably feel the same about my hammer. It's all what you get used to and what works for you. Still, there are some features you ought to consider carefully.

First, a professional hammer, regardless of the type of handle, has a head machined with a slightly convex surface, which stays on the head of a nail better than an ordinary hammer. Some framing hammers have a waffle pattern milled into the face of the hammer (see the photo below). The pattern gives the hammer a more tenacious grip on the nail compared to a smooth-face hammer. The down side of a waffle head is that it shouldn't be used when running any sort of trim, indoors or out: If you miss a nail with a waffle-head hammer, you'll see a very obvious bruise in the wood.

Although it's probably based more on psychology than physics, I prefer a straight-claw hammer to a curved-claw hammer. I think the straight claw puts the mass of the claw in a straight line with the striking face of the hammer to transfer more power to the nail

Suspenders on a tool belt can save your back by letting your shoulders carry most of the weight.

A framing hammer has one of two types of faces. The waffle face (left) grips the head of a nail more aggressively but leaves unmistakable damage if you miss the nail. The smooth face (right) is a better choice for finish work.

(at least that's my physics justification for my psychological preference). I also prefer the way a straight claw pulls nails compared to a curved claw.

I actually own six hammers and use them for different tasks. Three of them have long handles for framing. Because power is more important than accuracy in framing, a long handle means that the head is moving more quickly when it strikes the nail head. The longest of the three has an 18-in. handle, but my all-around favorite for framing is a 22-oz. waffle-head hammer with a 16-in. handle. It's light enough to swing all day long without making my arm tired and heavy enough to sink a 16d common nail with one or two well-placed swings. Heavier hammers are for tasks that require more force than finesse, such as tapping beams into place. The point is, find a hammer you like that fits the work you're doing.

Other Essential Hand Tools

A carpenter's tool belt is like a pair of comfortable old jeans. Some belts are made from wear-resistant (but heavy) leather, while others feature lightweight (but less durable) materials such as ballistic nylon. It's all what you're used to. I still wear the leather tool belt I bought back in the 1980s. Over the years, I've repaired holes in the pouches and bought a new hammer holster so that I wouldn't have to retire that belt, but the belt has served me well. One addition I made as I neared the tender age of 40 was a pair of suspenders to hold it up. Suspenders put the weight on my shoulders instead of on my aging lower back.

There are several hand tools that you should not be without while framing. You'll need a utility knife with a hollow handle for storing sharp blades. This tool comes in handy for sundry tasks from scoring a crosscut to sharpening carpenter pencils and opening cases of nails. Get one with a retractable blade, so the blade stays sharp and there are no surprises when you reach into your tool belt. In fact, get a couple: you're bound to lose at least one.

Another tool I keep in my tool belt is a pair of carpenter's pincers for pulling nails. Pincers are much less likely than a hammer to mar the smooth surface of a trim board. The wide, curved jaws are designed for superior grip and leverage when pulling nails. But when you need to dig out a nail buried deep in a 2×, a cat's paw is the only way to go. It's a highly effective tool, though it usually makes a mess of the board so use it sparingly and only where damage won't be visible.

Snapping chalklines is an essential part of any framing project, so I always carry a chalkline (or chalk box) in my tool belt. A chalkline consists of a reservoir that holds powdered chalk, along with a spindle of heavy twine. The powdered chalk coats the twine as it's pulled out of the box. When you've pulled out enough line to extend between measurement marks, pull the line as taut as you can, then lift the line straight up and

Carpenter's pincers are a must for pulling nails without marring finished surfaces. They're better than a hammer for this job.

A cat's paw is the best nail remover when the head of a nail is flush with the surface of the wood or sunk deeper.

release it (as shown in the photo on p. 36). You should be left with a laser-straight line of chalk on the surface if the line was taut enough. When you're finished snapping a line, mind the loose end to keep the twine clean and dry and to keep it from getting wound around things on the job site. There are many versions of the tool, each presumably a better mouse trap than the others. But I prefer my good old-fashioned model that

To snap a chalkline, anchor the hook and then stretch the line taut. Lift the line straight up a couple of inches, and let it snap back to the surface.

can double as a plumb bob in a pinch. However, one modern innovation I wish it had is a little door on the side to make adding chalk easier and less messy.

Levels

A good level is a must for ensuring that framing is installed perfectly plumb. Most levels are not adjustable, and it doesn't take much to render the vials inaccurate. Levels should be treated like the precision instruments that they are. If you have a case for your level, use it, and on the job site hang your level on a nail rather than leaning it against something where it's apt to be knocked over.

Some professional carpenters have as many levels as I have hammers. Levels come in many different lengths for specialized leveling tasks. The most common length is 4 ft., which is good for most framing tasks; 6-ft. levels come in handy for plumbing door openings. And many companies make levels that telescope to lengths of 10 ft. to 12 ft. to plumb tall walls, such as gables.

Most carpenters carry a short level called a torpedo level in their tool belts. Because of its short length, this level is usually reserved for quick approximations, not extreme precision, or for locations where a longer level just wouldn't fit. Torpedo levels are handy for leveling

Chalk Choices

I always keep two types of chalk on hand while framing a house: nonpermanent and permanent. They're color coded to distinguish them. The most common nonpermanent chalk color is blue. Use blue chalk for lines that might have to be adjusted later, because you can easily erase them. Blue lines wash away in the lightest rain, however, so if you're snapping lines that need to stand up to weather and foot traffic, use a permanent chalk, commonly red. Keep a separate chalk box for each color.

stair stringers and treads, for example. But a torpedo is the wrong level for some tasks. If you put it against the top of a bowed or warped stud, for example, it will read out of level one direction, but put it at the bottom and it will give a different reading. You can hold a torpedo level against a longer straightedge to make it more effective in a pinch, but at that point you might as well reach for a longer level if one is handy.

ESSENTIAL TECHNIQUE

Checking Your Level

"Good enough" is not an option when checking walls for plumb. Levels can be knocked out of whack during the rigors of framing a house, so check the accuracy of your level often. The easiest and quickest way to check a level is to put it up against a known or presumed plumb surface such as a door jamb on a finished house. Check the vial and note what it reads. Then, keeping the level in the same location, flip the level so that the opposite edge is against the jamb. The reading should be exactly the same.

If the surface you're using is slightly out of plumb, try this method of testing. Slip a shim behind the level and slide the shim down until the level reads perfectly plumb. Draw a line on the shim at the end of the level and then flip the level to the opposite edge. It should read perfectly plumb when the mark on the shim is still lined up. If the level is off, even by a little, put tape across the vial so you won't be tempted to use it. Test all the vials that way and mark on the level which vials are trustworthy.

To calibrate a level, put it against a surface such as a door jamb. Shim the level until it reads plumb and mark the top of the level on the shim. Now flip the level and align it with the mark, then check the vial. If the vial still reads plumb, it can be trusted.

my living as a carpenter, and that's when I bought my first professional-quality circular saw. I still have that saw, and it still works well; it just feels good in my hands. If you're planning to frame your own home, invest in a good saw—it will pay dividends in accuracy and enjoyment.

Sidewinders versus worm-drive saws

My circular saw is the type known as a sidewinder. The most common version uses a 7¼-in.-dia. blade and is fairly light in weight. But there are other saws. In the late 1980s, I worked on a couple of timber-frame projects. I needed a saw with much more torque to chew through the timbers, so I bought an 8¼-in. worm-drive saw that could cut through red oak at a maximum depth without bogging down. It was years before I realized that carpenters tend to differ in their saw preferences depending on where they live.

West Coast framers use worm-drive saws almost exclusively and consider sidewinders to be toys. The torque of a worm-drive makes it easy to gang-cut stacked lumber, a faster technique than cutting boards one by one. But using a worm-drive all day long is hard work because the tool is just plain heavy. Framers skilled at using a worm-drive saw can use that weight to their advantage when cutting, but veteran wormies will be the ones wearing forearm braces down the road. East Coast framers, on the other hand, prefer sidewinders so that's what you'll see in this book.

Saw manufacturers seem to change their circular saws like most people change socks. There always seems to be a new gadget or gimmick to make the newest saw a little better than last year's model. Look beyond the gimmicks when choosing your saw. Read tool reviews in trade magazines and online to see which tools are recommended for longevity, ease of use, and safety. Then go to a tool dealer that carries several different brands. Most professional tool dealers will let you try out a saw at the store.

Be aware of the weight, the size of the handle, and the ease with which you can access the trigger. I know guys with big hands who can't get their fingers on the trigger of some saws, especially with gloves on. Check

A circular saw is a framer's workhorse. It is lightweight, easy to use, and can be locked at an angle for bevel cuts.

Choosing and Adjusting a Circular Saw

My earliest recollection of carpentry is of my dad finishing the attic in our rambling Rhode Island ranch house. He used only hand tools and was a sucker for unfinished lauan mahogany paneling. No drywall for him, no sirree. Looking back, I can remember him sweating as he sawed 2×s by hand.

I bought my first circular saw when I was in college to help with building display bases for sculptures. It worked okay. But in 1980, I decided to make to see how easily you can hold the saw while holding the guard in a raised position, a maneuver you must be able to do when cutting steep angles. If you have small hands, difficulty with this feature can be a deal breaker. By the way, you should *never* remove the guard or clamp it in a permanent retracted position.

After that, look at features such as cord length and amperage. At this point, I would not be tempted to buy a cordless circular saw for constant use because of battery life. A longer cord is helpful, but 6 ft. is usually more than enough. Many companies tout the amperage of the saw motors as a selling point. But if you run the saw off long extension cords, higher amp motors are more likely to trip the job-site circuit.

A good blade is a must

To make any circular saw work at its best, fit it with a good sharp blade. And make sure the blade you get is for framing, not finish work. Most manufacturers label their blades as such. Framing blades typically have 24 teeth set at a positive hook angle for fast, aggressive cutting. You might think that more teeth make a better blade, but when it comes to cutting framing lumber, extra teeth just slow things down.

Most framing blades for circular saws are thin-kerf blades, meaning that the width of the teeth is relatively narrow. A thin-kerf blade removes less wood so it cuts more quickly, an advantage in framing. However, thin-kerf blades may not be the best choice for more precise finish work, where speed is less important than a smooth cut.

Steel sawteeth were once common, but these days hardly anyone bothers with them. The best blades have a carbide tip on each tooth that stays sharp far longer than steel. In the course of framing a moderate-size house, you can expect to run

through four to six good blades. Carbide chips if it hits something hard, such as an embedded nail or screw, and even carbide wears down eventually. When you feel yourself having to apply a lot of pressure to keep the saw moving through a board, or if the saw starts to bog down during cuts or consistently pulls to right or left, stop and put in a new blade. If a blade is dull but the carbide teeth are still good, I send it out to be resharpened, but always have a couple of sharp blades on hand to replace it.

Most carbide-tooth blades meet their demise when the carbide on one or more teeth chips or breaks off completely. There are outfits that can recarbide teeth or machine all the teeth down to match a chipped tooth, but those options typically cost more than the blade is worth.

Basic saw adjustments

One of the nice features of a circular saw is that relatively few adjustments are needed in the course of a day. But those few adjustments are critical for safety and performance, so practice them a bit before you tackle a project.

Good cuts start with good blades. These blades are designed specifically for framing work. They have thin-kerf, aggressive carbide teeth. Always keep a sharp blade in your saw for fast, accurate, and safe cuts.

Changing a blade The blade in a circular saw is always mounted so the teeth face toward the front underside of the saw's table. In that configuration, the action of the blade pushes the stock toward the base. Every saw has an arrow (usually on the guard) to indicate the proper direction of the teeth.

When changing the blade, *always* unplug the saw before you start. Most saws have a shaft lock that keeps the blade from turning as you loosen or tighten the bolt that holds it in place. Engage the shaft lock, then loosen the bolt carefully. Always make sure the saw baseplate is sitting securely on a flat surface and always push the wrench down and away from the blade teeth rather than pulling the wrench up and toward the teeth: If the wrench slips off the head of the bolt, you don't want your hand to hit the blade, even a dull one.

With most circular saws, the bolt that secures the blade tightens in the opposite direction of the blade rotation, and every saw I know has the blade direction stamped on the housing as a reminder of this.

Squaring the blade to the saw table One last check I always make before starting a cut is to make sure that the blade is perpendicular to the saw's baseplate. If the blade is tilted even slightly one way or the other, your cuts won't be square and the joints won't be tight.

To square the blade, first unplug the saw and then adjust the cutting depth so the blade is at its lowest (most exposed) position. Turn the saw upside down, retract the blade guard, and rest a triangular square against the blade as shown in the top photo on the facing page. If there's a light background behind the saw, you'll easily see any gap between the blade and the square. Adjust the angle of the baseplate until the gap disappears. When the blade is set, make a test cut and check to make sure that it's square. Many saws have a set screw or a stop that lets you return the blade to exactly 90 degrees after you've cut at an angle. Once you've established that the blade is square, set the stop so that you don't have to square the blade every time you change cutting angles.

Setting the blade depth Circular saws cut more smoothly and safely when the depth of the blade is set just slightly greater (about ¼ in.) than the thickness of the stock. If too much blade is exposed below the stock, you just might cut through more than you had in mind. But an overexposed blade also increases the chance that the saw will bind and kick back at you (see "Preventing Kickback" on the facing page). Kickback is a hazard created when the blade of a circular saw twists or gets pinched in the midst of a cut. When kickback occurs, the saw jerks away from the work with such a sudden and violent force that you lose control of the saw before the blade has time to stop. You can imagine the consequences.

To provide an extra measure of protection against kickback, always position yourself beside the saw as you cut; never stand directly behind it.

The easiest way to set the blade depth is to place the saw directly on the stock with the guard pulled up, so that the blade is exposed alongside the stock. Next, release the lever that locks the height adjustment and lower the blade until the proper amount is exposed. Tighten the lever, and you're almost ready to start cutting.

Basic Saw Techniques

A circular saw might seem intimidating the first time you pick one up. That's understandable, but with continued use you'll come to see it as an essential framing tool. Don't ever treat it casually or use it carelessly, because a circular saw injury is rarely insignificant.

If you wear prescription glasses, you can have safety glasses made for your prescription or find models that fit comfortably over your regular glasses. These days, you'll find safety glasses in many styles to fit any face comfortably. I count them as cheap insurance considering the alternative—losing my sight. By the way, standard sunglasses aren't safety glasses.

To set the blade square to the baseplate, place a triangular square against the blade and then look for gaps between them.

Preventing Kickback

- Whenever you crosscut with a circular saw, support the board so that the saw kerf won't close up as you make your cut and pinch the blade.
- If the board is supported on either side of the cut, keep the board as straight as possible until you've completed the cut.
- If the board is supported to one side of the cut, make sure the waste can fall away cleanly after you've cut through.
- Make sure the blade is set to the proper depth and operate the saw while standing to one side of the saw, not directly behind it.
- Make sure the work is held securely at all times so it doesn't shift as you cut, twisting the blade into the wood.
- If necessary, use clamps to secure small pieces while they're being cut.

For safety and ease of cut, unplug the saw, retract the guard, and set the blade depth just beyond the thickness of the material you're cutting.

The easiest and most efficient way to cut stock is when it's still in the stack. The stack then becomes your workbench.

You need the real deal, a National Institute of Occupational Safety and Health (NIOSH) approved pair that offers serious protection.

Making a crosscut

Most of the cuts you'll make in the course of framing a house are crosscuts. These are cuts made across the grain of the wood. A cut made perpendicular to the edge of a board is a 90-degree crosscut, and, believe

me, you'll make a lot of these when cutting boards to length. If you think of the grain as a tightly compacted collection of string, a crosscut chops through the string like a knife or scissors.

Quite often in framing you can crosscut stock directly off the pile or "clip" (the factory-bundled package) of stock. Measure and mark the cut, then slide the board so that the waste end is hanging off the pile. By doing so, the weight of the waste pulls it naturally away from the blade. To make the cut, put the blade on the proper side of your layout line with the blade 1 in. or so away from the wood. Then pull the trigger, let the saw come up to speed, and cut straight through the wood using firm, steady pressure. Push the saw through the board until the waste drops free. Release the trigger and set up for the next cut.

Most framers use only one hand to push the saw and use the other to secure the stock, keeping it from slipping out of position. If the stock slides on the pile, you may find yourself cutting into other boards, or even worse, the saw may kick back and hurt you.

When you make a standard crosscut, the blade guard lifts automatically as the cut begins and then

Guiding a Crosscut

A triangular square works well as a saw guide when making square crosscuts. Align the sawblade with the line or mark and push the square up against the saw table and square to the edge of the board. Firmly grip both the board and the square to prevent them from shifting as you make the cut.

closes over the blade when the cut is complete. However, the guard doesn't easily retract if the saw enters the wood at a steep angle. That's also the case with a bevel cut, a type of crosscut made with the sawblade *not* at 90 degrees to the baseplate. (You'll get plenty of practice with bevel cuts when it comes time for roof framing.) To prevent the saw from getting jammed as a bevel cut starts, you have to manually retract the guard until the blade is well engaged. Before starting the cut, rotate the guard lever up with your left hand, then pull the trigger and push the saw into the wood with your right hand (see the photo at right). Once the saw gets a good bite on the wood, you can release the guard lever. With both hands working the saw it's often hard to hold the board steady, so pin it down with a knee or a clamp, or have a helper hold it steady as you make your cut.

Sooner or later a saw will get hung up in a cut for any number of reasons. The saw will suddenly feel sluggish and you'll hear a difference in its sound. Your first reaction will be to pull the saw backward slightly to take another run at the cut. *Don't do it!* Instead, release the trigger immediately and hold the saw in place until the blade stops. Pulling the saw backward

while the blade is spinning is just asking for kickback (see "Preventing Kickback" on p. 41).

Making a rip cut

A cut made parallel to the grain or edge of a board is called a rip cut. One example would be ripping a 1×6 into a 1×4. Rips are made along the length of a board so they are typically longer than crosscuts. There are

To keep the blade guard from hanging up during angle cuts, hold the guard up with one hand while squeezing the trigger and pushing the saw with the other.

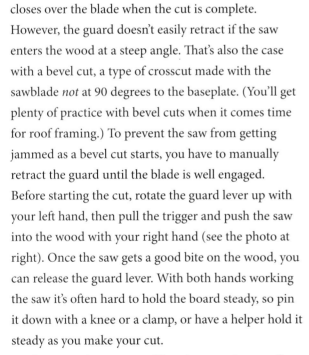

To make a rip cut quickly and efficiently, use a rip guide that rides against the edge of the board (top). If a guide isn't available, you can use locking pliers (middle) or your finger (bottom) as a rip guide.

When cutting sheet goods, working directly on the pile is fast and efficient. A 2× separates the sheet being cut from the rest of the pile.

A board clamped at either end of the sheet acts as a guide for making long, uniform cuts.

several ways to guide the saw while making a rip cut.

The first method is to use the rip guide (sometimes called a rip fence) that comes with the saw. Line the saw up with the cut line, then slide the guide to the edge of the board and secure it to the saw by tightening the thumbscrew. Let the guide ride along the edge of the board as you make your cut. Be careful not to let the saw pivot on the guide, especially at the end of the cut, which would likely result in kickback.

Another way to guide the saw for thin rips is attaching locking pliers to the saw table. Likewise, your finger can make a decent rip guide in a pinch.

Cutting sheet goods and trim

To build a house you'll cut a lot of sheathing. When cutting sheet goods, it's often easiest to work off the pile. First be sure the blade is at the proper depth for the sheathing that you're cutting. Place a 2× between the sheet you're cutting and the sheet below to separate them. For long, narrow cuts you may need a rip guide, or you can clamp a guide board to the sheet. For most sheathing cuts, however, you probably won't need a guide if you can follow the line reasonably well.

Minimizing tearout on trim A circular saw actually cuts from the bottom of the board to the top, so the cleanest cut is usually on the bottom side. The top edges where the blade is exiting are more likely to be rough or tear out. For most framing, tearout is not something to worry about. However, if you use a circular saw to cut trim boards such as rakes and fascias, where the cut will forever be visible, tearout can render a joint unacceptable.

There are two ways to minimize tearout. The first is to cut from the back side of the board. In this case, the cut line needs to be transferred to the back so you can cut with that side facing up. Another method is to score the cut with a utility knife before making the cut, as shown in the photo above. Scoring actually

precuts the top of the board so the blade doesn't lift the grain as it exits.

To minimize tearout, score the line with a utility knife before cutting.

Nailing

I had my first experience with house framing in the early 1980s. We nailed those frames together entirely by hand. Just a few years later, I found myself working in the joinery shop at a boatyard. The majority of the fastening tools in that shop were pneumatic, and I soon learned how versatile air-powered tools could be. A few years later still, I was back building homes and found that pneumatic nailers had become common. In the early 1990s I took the plunge and bought a compressor, a framing nailer, and a finish nailer.

Compressors are indeed a blessing, but they can also be a pain in the butt. They're usually very noisy, which can be annoying when crew members try to communicate. They draw a lot of electricity, especially during startup, which can trip the job-site circuit breaker if any other electricity is being used at the same time. And they don't start easily in cold weather. But despite those faults, compressors are worth having because pneumatic nailers have become indispensable for framing quickly and efficiently with less physical strain. The crew that built the house featured in this book used pneumatics almost exclusively. If you're planning to frame your own house, seriously consider buying a compressor, a framing nailer, and enough hose to get the job done. You won't use other nailers nearly as much, so you can always rent what you need. Don't throw away your faithful old hammer; just plan on using it a lot less.

The compressor and hoses

Compressors come in many shapes and sizes, from tiny portables that can run only a single finish nailer to massive shop compressors that can run dozens of tools simultaneously. Framing nailers need a lot of air at fairly high pressure, so for a small crew, the compressor ought to be big enough to supply two or three nailers without having to run constantly. But it also should be small enough that you can get it in and out of your truck or van easily.

Most compressors have two pressure regulators: one that turns the compressor on and off and one that regulates the amount of pressure going to the hoses. The first regulator is set by the factory so that the pressure in the tank does not exceed safe limits. The second is set by the crew according to the task at hand. For example, assembling a 2× frame requires fairly high pressure, and it doesn't matter if the nailer drives the nails slightly below the surface of the wood. But with sheathing, code specifically states that nails must not be overdriven. The head of a nail driven too deep breaks the outer layers of the surrounding wood, which substantially reduces the holding power of the nail. So the pressure for the sheathing nailer should be set a lot lower than for a framing nailer.

Hoses come in a wide variety of shapes and sizes. I bought heavy rubber hoses with my compressor, but they're hard to lug around, and they get stiff in cold weather. The hoses seen in this book are thin and lightweight, plus they coil up very easily at the end of the day. They also seem to last as long as the rubber hoses.

Basic maintenance and care
One note of caution when handling hoses: Don't let the ends of the hose get dirty. Any dirt or dust that gets into the hose will get into a nailer and ruin it. It's also a good idea to add one drop of compressor oil to the air intake of a nailer at the beginning of every day to keep the inner workings lubricated and functioning properly.

A compressor has two basic controls: an on–off lever (red tip) and a pressure regulator that controls the amount of air pressure dispensed to the hose.

Long Hose or Long Extension Cord?

On most of the job sites I've worked on, the compressor was parked close to the house and ran off of a long extension cord plugged into a receptacle. Turn on a saw while the compressor was powering up and you'd trip the circuit breaker. But on this project, the compressor power cord was plugged directly into a receptacle at the house next door (with the owner's permission) and a long feeder hose ran to the building site, where it was fitted with a manifold to distribute air to multiple tools. The result was a quieter job site, and not once during framing did a circuit breaker have to be reset.

Keeping the compressor removed from the site means a quieter site and less electrical demand for running the compressor. Here the compressor is kept at a nearby home and connected to the site via a long hose.

At the end of each day, turn off the compressor, bleed off the air, and drain any moisture that has condensed in the tank. A compressor's biggest enemy is moisture that accumulates inside the tank. If not drained regularly, the tank will rust and eventually fail.

Types of nailers On this house, we used at least a half dozen different nailers, but the workhorses were the framing nailers. Framing nailers come in two different styles: Those that accept nails collated in a straight line, or stick, and those that accept nails collated in a coil.

Framers typically use stick nailers for joining framing members because they can drive a wide variety of nails. For sheathing, they switch to a coil nailer. The coil magazines hold more nails, so they don't need to be reloaded as often. When nailing off a large expanse of sheathing, you don't want to have to stop and reload repeatedly. Whatever nailer you choose, be sure to use nails specified for that nailer. Some nailers drive nails only with clipped heads, and most nailers require a specific collation medium (the stuff that holds the nails together) such as paper or wire.

Installing metal connectors was once a dreaded chore because pneumatic nailers just weren't accurate enough for the work. As a result, all those short, stubby connector nails had to be driven by hand through holes in the connector flanges. No more. A positive placement nailer can drive a nail exactly where you want it and makes that job easy. The nails are specifically designed with unusually high shear strength, making them ideal for securing metal connectors. (Don't ever use standard nails to secure a framing connector.)

Pneumatic nailers come in many types and configurations. A stick nailer (above left) carries headed nails that are collated in a straight line. A coil nailer (above right) carries more nails.

Another nailer used by framers is the medium-gauge finish nailer. This nailer drives finish nails of various lengths. On the project house, for example, we used stainless-steel finish nails to attach exterior trim. We also used a coil nailer that drives smaller-gauge headed nails to attach corner boards and fascia.

Nailing with Pneumatics

Having sung their praises, I must admit that pneumatic nailers are the cause of many job-site injuries. Here are some essential safety guidelines:

- Never fire the nailer toward yourself or in the direction of another worker.

- Never place your hands in the line of fire. If you're holding a board in place, keep your hand back a safe distance (at least 1 ft. away). Otherwise, if a nail hits a knot or another nail, it can deflect, come out through the side of the board, and sink into your hand.

- Never keep your finger on the trigger except when you are ready to fire the nailer.

Many nailers are equipped with settings for bounce fire or single fire. On the bounce-fire setting, you can keep the trigger pulled and the nailer fires every time the nose is pressed against something solid. Framers sometimes do this because it is fast. However, on the single-fire setting, the trigger has to be released and squeezed again before the nailer will fire. The single-shot setting is much safer, and novices should use it for all nailing tasks.

Tolerances: How Close Is Close Enough?

I've never been asked to work to more exacting tolerances than when I was a boat carpenter. Joints had to be fitted within $\frac{1}{64}$ in. That's one quarter of a 16th—furniture tolerances! Frame a house to those tolerances, and you'll be old before you get the roof on. Of course, frame to tolerances at the opposite extreme, and maybe the roof won't stay on. So it's a subject that inspires considerable debate. Production framers, who have been known to frame an entire house in a day, actually leave a certain amount of slop in joints to make the job go more quickly. Custom framers who pride themselves on tight, well-fitting joinery make every joint as perfect as possible. I like to land somewhere in between the two.

The three most important words in discussing tolerances are *square, plumb,* and *level.* Square means that every corner—horizontal or vertical—is exactly 90 degrees, unless the plans specify otherwise. Plumb means that all vertical surfaces and assemblies such as walls are parallel with the forces of gravity, and level means that all horizontal surfaces are perpendicular to the forces of gravity. The three concepts are interrelated. If you build a square wall on a level deck and raise it to a plumb position along its length, the wall should also be plumb at the ends, and the top of that wall should be level. Conversely, if the wall isn't square or the deck isn't level, plumb and level aren't about to happen.

I strive for making a house as square as possible. That means checking the footprint at every crucial step to make sure it's square, and making adjustments if it isn't. Doing so makes my job as framer easier as I go along, but it also helps the guy who is going to do the finish work. Throughout the book I'll reemphasize this point.

When checking for square using long diagonals such as for the first floor of the house, measurements should be within ⅛ in. The smaller the square, the smaller the tolerances. For example, when checking diagonals on the garage sills, I'd aim for tolerances of

¹⁄₁₆ in. The trick is to tweak things as you go along. Don't spend an hour getting the mudsills absolutely perfect, because chances are you'll still need to do more adjusting when you're ready to sheathe the first floor.

While tolerances for plumb should be near perfect, the framing to reach those tolerances doesn't have to be. For instance, where the second top plate overlaps the first, a little play on either side of the board won't make a difference in the integrity of the house as long as the wall is installed plumb. Accepting lesser tolerances where they're appropriate lets you frame the house much more quickly and efficiently. And often framing to lesser tolerances gives you more latitude for adjustment if necessary.

A Word about Weather

It's ironic that one reason we build a house is to protect us from the weather, but there's little to protect us from weather as we build. In fact, weather is the most formidable enemy you'll face while framing. Here in the Northeast, framing in the winter often means scraping ice and snow off of lumber before you can use it. It can also subject the inside of the house to weeks of moisture that must dry out before the finish stages of the house can begin. And working in bad weather is less safe and just plain takes longer. A wet chalkline is useless, and wiping moisture off the blade of your tape after every measurement slows work considerably. Whether you're having to mark and cut wet lumber or just move around a muddy, slippery site, quite often it's better to just wait for a better day. Plan to frame your house in the best weather your area offers.

Every framer I know is an avid weather watcher. Stacks of lumber are always covered with tarps if bad weather is on the way or not. Call me superstitious, but leaving lumber uncovered is an invitation to the winds of

fate to bring in that unpredicted downpour or snow squall. Depending on what is completed inside the house, I've known framers who cover a sheathed roof or floor with large tarp to keep the rain or snow out. If a wall or gable is raised without another wall to tie into, extra bracing should be added to resist wind—raising a single wall should not be the last thing you do at the end of the day.

The operative words here: *common sense.* Think ahead and make sure you don't leave any parts of the house vulnerable if you can avoid it.

Weather can sometimes curtail work on the job site altogether. Extra care has to be taken when wood is wet and slippery, and in steady rain or snow, measurement marks and snapped lines can disappear quickly.

3

Framing the First-Floor Deck

The most important part of a house frame is the first-floor deck: the mudsills, the beams that support the floor framing, and the floor frame itself, including the joists and sheathing. If you want the rest of the house to be level, square, and plumb, you'd better start with a square and level platform.

Before getting started, look over whatever plan shows the floor-framing details; they might be on the first-floor framing plan or, as in this case, on the foundation plan (see "First-Floor Framing Plan" on the facing page). Make note of any special details, such as openings for stairs and chimneys, and look for areas marked for extra framing needed to carry heavy loads, such as large bath tubs. Pay particular attention to locations

First-Floor Framing Plan

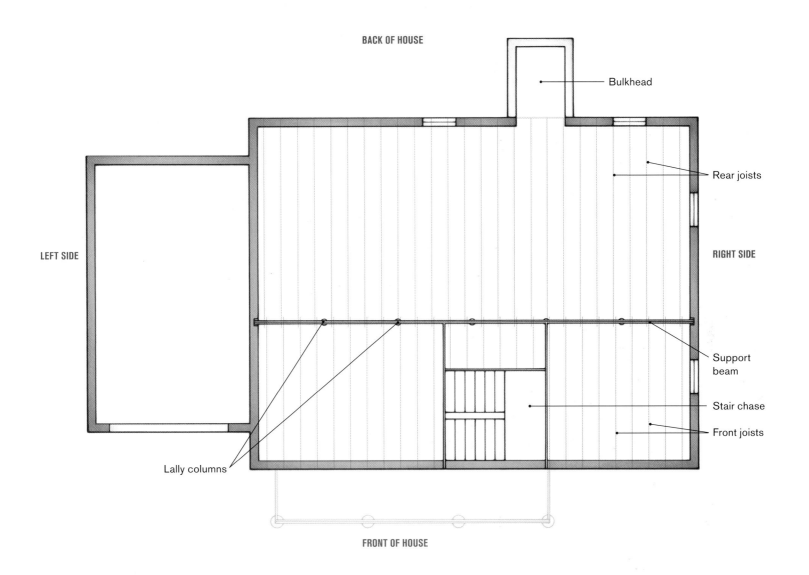

BACK OF HOUSE

Bulkhead

Rear joists

LEFT SIDE

RIGHT SIDE

Support beam

Stair chase

Front joists

Lally columns

FRONT OF HOUSE

where the joists change direction or cantilever over the foundation walls to support a bay window or balcony. These areas are easy to overlook as framing progresses and, believe me, you'll want to get them right the first time. Use a highlighter to flag these features so you'll be sure to pay attention to them as framing progresses.

Over the years, I've been lucky to have worked for builders who were meticulous about getting the first-floor deck right, but I also spent a short time on a crew where speed trumped accuracy. Walls on the shoddy first-floor deck seemed to go up easily . . . until we tried to straighten and square the second floor.

Walls had to be set out of plumb or off the layout to compensate for discrepancies in the first-floor deck. At the roof, the rafters didn't meet the ridge squarely, and with the framing out of whack, every piece of fascia had to be custom cut to fit. Trimming out soffits and returns that weren't square was an even bigger headache.

My advice: Take your time and get the first-floor framing as close to perfect as you can. Otherwise, you'll have to spend extra time and money to fix problems later. Get this part right, and the house will be better for your efforts.

Check the foundation for square by measuring to the outer corners in both directions. There should be less than 1-in. difference between the measurements.

Mudsills

Once the foundation is complete, framing starts with the mudsills, sometimes called sills or sill plates. Mudsills are made of rot-resistant, pressure-treated wood and are bolted to the top of the foundation. Historically, however, mudsills literally sat in the dirt below the house. If you want to get floor framing right, squaring and leveling the mudsills is a critical first step.

Check the foundation

The first step is to measure the foundation and give it a quick check for square. Don't assume that your foundation contractor got it right, because even the best have bad days. To check a foundation for square, have two crew members stretch a 100-ft. tape diagonally from one corner to another, as shown in the photo above. Note the measurement and repeat the process at the other two corners. The numbers don't have to match exactly but they should be less than

1 in. apart. If the difference is greater, you can sometimes compensate by installing a wider mudsill and letting it hang over the foundation slightly, but this arrangement is hardly ideal.

In addition to taking diagonal measurements, I also measure between the longest parallel walls of the foundation to make sure they are indeed parallel. It's also a good idea to look for irregularities in the sides or top of the walls. If they bow outward, for example, you might have to position the mudsills so that the wall sheathing will clear the bowed area. If there's a hump or belly in the top surface of the foundation, you may have to scribe or shim the mudsill. If you have a transit on site, check the top surface of the foundation in various locations to make sure it's fairly level (see "Using a Transit" on p. 54). If you don't have a transit, you can use a water level—a long, clear length of plastic tubing filled with water—but it's not nearly as quick and convenient as using a transit. A foundation should be within ⅜ in. of level. If it's greater than that, one solution is to install a second mudsill over the first and shim it to level (see "Double-Layer Mudsills" on p. 60).

STEP BY STEP

Framing the First-Floor Deck

1 Check the foundation for square.
2 Snap chalklines for mudsill alignment.
3 Install mudsills.
4 Build and install support beam.
5 Lay out joist spacing.
6 Install joists and rims.
7 Frame openings in the deck.
8 Install floor sheathing.

ESSENTIAL TECHNIQUE

Taking Long Measurements

To take long measurements, such as the diagonals used to square a foundation, one crew member holds the end of the tape while another crew member reads the measurement. Because it's difficult to hold the hook of the tape at a specific point, the person on the "dummy end" usually holds the tape at a given mark such as 1 in. The person at the other end then subtracts that inch from the overall measurement, a procedure called *burning an inch.* But many carpenters burn a foot instead because subtracting a foot is easier. Either way, the person holding the end of the tape should call out the number he's holding ("Burning a foot!"). The person reading the measurement should then acknowledge that number.

After measuring and marking the width of the mudsill at each end of the foundation (top), snap a chalkline between the marks (above). All other wall locations spring from this baseline.

Snap lines for parallel walls

Once you've established that the foundation is reasonably square and level, it's time to mark out the exact location for the mudsills. This house has 2×6 sills, so the first step is to measure 5½ in. (the actual width of a 2×6) inward from the outside face of the foundation. Make your measurements at each end of the longest wall and mark those points. Then snap a chalkline between the marks. This line will serve as the baseline from which all subsequent measurements will be taken. Snap the lines in nonpermanent blue chalk so you can adjust them later if necessary.

Next, move to the wall parallel to the baseline and make a mark 5½ in. from the outside of the founda-tion at one end. Measure the distance from the baseline to that mark. At the other end of the foundation, mark the same distance from the baseline (see the top photo on p. 54), and then snap a chalkline between those marks. Now you've established a layout line for the wall that's parallel to the baseline.

Snap lines for perpendicular walls

The next step is to mark perpendicular layout lines for the mudsills on the end walls. The most common method for establishing perpendicular lines uses the Pythagorean theorem (remember high school alge-bra?): $A^2 + B^2 = C^2$. A triangle with those proportions always has one 90-degree angle. Carpenters refer to

To establish a layout line for the parallel mudsill, measure and mark the same distance from the baseline at each end of the foundation, then snap a chalkline between the marks.

Baseline wall

Parallel wall

Using a Transit

A transit is a surveyor's tool that builders have adopted for checking level over long distances. It's a precise optical tool (though essentially just a telescope atop a tripod). To use a transit, one crew member sights through the telescope while another holds a measuring tape or a rod (a stick with graduated measurements) at various locations on the foundation. As the crew member with the tape moves from place to place, the other crew member sights and records measurements. If measurements at two different points are equal, then those two points are level in relation to each other. A transit is very useful, especially for the early stages of framing. It's an expensive tool but can usually be rented.

A transit can be used to determine how level the foundation is. Sight through the telescope as a helper moves a tape or measuring rod along the foundation or mudsill.

the simplest of these triangles as a 3-4-5 triangle because any multiple of those numbers yields a right triangle with an easily determined hypotenuse. Here's the basic example: $3^2 + 4^2 = 5^2$ (9 + 16 = 25). If you wanted to work with a slightly bigger triangle, 3-4-5 multiplied by three yields a triangle with a 9-ft. leg, a 12-ft. leg, and a 15-ft. hypotenuse. The larger the triangle, the more accurate the layout will be. Carpenters rely on 3-4-5 triangles to establish or verify perpendicular lines at many stages in the framing process. In practice, determine the length and position of one leg and the corner where you need the 90-degree angle, then figure out the length of the other leg and the hypotenuse. Piece of cake.

ESSENTIAL TECHNIQUE

Snapping Long Lines

To snap a chalkline, pull it taut, lift it straight up, and then let go. Snapping a long chalkline this way, however, often results in a line that's not perfectly straight. A better approach is to have someone go to the middle of the stretched line, hold it in place with a thumb, and then snap the line on both sides with the other hand. This technique is particularly useful on windy days and when the surface is somewhat irregular.

The 3-4-5 Triangle

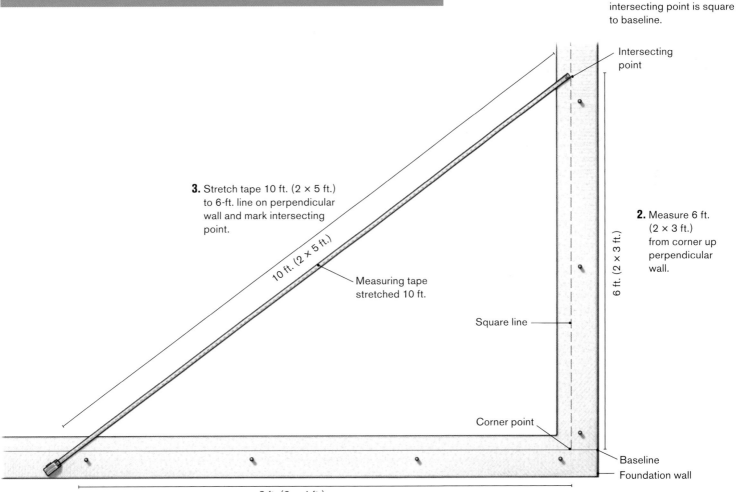

4. Line from corner through intersecting point is square to baseline.

Intersecting point

3. Stretch tape 10 ft. (2 × 5 ft.) to 6-ft. line on perpendicular wall and mark intersecting point.

10 ft. (2 × 5 ft.)

Measuring tape stretched 10 ft.

2. Measure 6 ft. (2 × 3 ft.) from corner up perpendicular wall.

6 ft. (2 × 3 ft.)

Square line

Corner point

Baseline

Foundation wall

8 ft. (2 × 4 ft.)

1. Measure 8 ft. (2 × 4 ft.) from corner point on baseline.

One way to locate the perpendicular mudsill is to lay out a 3-4-5 triangle (right). Extending one leg of the triangle along the foundation creates a line that's square to the baseline (above).

Here's how it worked on this project. To lay out the mudsills, we measured 5½ in. along the baseline from one end and marked a point that represented the inside corner of the intersecting mudsills. Then we measured along the baseline 8 ft. (a multiple of 4) from that mark and made a second mark. From that point we measured diagonally exactly 10 ft. (a multiple of 5) toward the intersecting wall and jockeyed the tape back and forth until it matched the 6-ft. point (a multiple of 3) on a second tape stretched from the inside corner mark. We made a third mark at this

intersection point. Finally, we snapped a chalkline from the baseline corner, through the third mark and all the way to the parallel foundation wall to establish the line for the perpendicular sill.

To locate the layout for the second perpendicular mudsill, we first measured in 5½ in. at the other end of the baseline and marked the other corner point. Then we measured the length between the baseline corners. Next we measured that same distance along the parallel mudsill from the corner just established and marked the final corner point. A chalkline snapped between the remaining corner points is the layout line for the second perpendicular wall.

As a final check, take diagonal measurements between corner points as you did when checking the foundation. At this point, the diagonals should be within ⅛ in. of each other. If they're not, go back through the procedure and adjust layout until the diagonals match. If any changes are required, snap new chalklines in a different color chalk to avoid confusion. Once the layout for the house's mudsills is complete, lines for the garage can be figured using the same simple geometry.

With the first perpendicular mudsill located, measure the length between the corners of the baseline mudsill and mark that length on the parallel mudsill line, giving you the fourth corner.

Laying Out Intersecting Sills

Rhode Island builders Rick Arnold and Mike Guertin use a different method to lay out the mudsills and think it's simpler and more accurate; try both methods and decide for yourself. After establishing a baseline on one wall, stretch a tape from each of the inside corner points of the mudsill to a point near the middle of the parallel wall. Cross the tapes and adjust them until the measurements are equal. This point is the exact middle of the parallel wall.

Divide the distance between the baseline corners in half. Then measure and mark the halved distance from either side of the parallel wall's center point. Those marks are the corners for the perpendicular mudsills. As with the 3-4-5 method, use diagonal measurements between the corners to confirm that the sill lines are square.

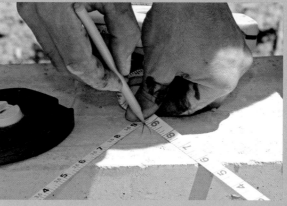

Stretch two tapes from the baseline corners to the middle of the parallel mudsill. The exact middle of the parallel mudsill is where the tapes intersect at the same measurement.

ANOTHER WAY TO DO IT

Installing Sill Sealer

Recently a builder showed me an installation technique that was a real head slapper. After drilling anchor bolt holes he flipped the mudsills upside down and attached the sill seal with a few staples. With this method, the sill seal stays perfectly aligned with the sill stock, and it can't blow or slide around before you get the sills in place.

If you won't be able to install the mudsills right away, go back and snap all the lines in permanent chalk (see "Chalk Choices" on p. 36).

Cut and install mudsills

Once the sill layout is complete, you're ready to cut and install the mudsills, a process that starts with the installation of sill sealer. This is a thin, flat insulating foam that plugs minor gaps and irregularities between the mudsills and the foundation to prevent cold air from leaking into the house. The usual installation method isn't complex. You just roll out lengths of sill sealer and impale it on the anchor bolts. The inside edge of the stuff should be aligned with your chalklines.

With sill sealer in place, you can begin to cut the mudsills to length. You can take measurements directly from the lines on the foundation to determine the lengths of all the mudsill stock, but it's actually faster and more accurate to cut and fit the boards one by one. Regardless of the method you choose, always make sure the ends of the mudsills are square so that the boards butt together nicely. When you get to a corner, let the board run past the edge of the foundation—it's easier to cut the sill to length after it's in place.

As each mudsill is cut to length, it must be drilled to fit over the foundation bolts. There are many ways to do this, so I'll explain the method I see most often. Place a length of mudsill stock on the outside of the foundation wall, as close to the bolts as possible and parallel to the chalkline. Measure from the chalkline to the *center* of a bolt (see the drawing on the facing page). Then transfer that measurement to the mudsill, measuring from its inside edge, and make a mark. Now hook your square on the edge of the mudsill, with the corner lined up to the middle of the bolt, and draw a line to the mark. The intersection of the line and the mark is the center point for the bolt hole. When you've marked the bolt locations, drill a *slightly* oversize hole at each mark. The extra room will let you tweak the mudsill to fit perfectly along the layout line on the foundation.

Marking Bolt Locations

1. Measure from snapped line to center of bolt (A).

2. Measure same distance from edge of sill stock.

3. Align rafter square with center of bolt, and mark where edge of square intersects with measurement for center of bolt hole.

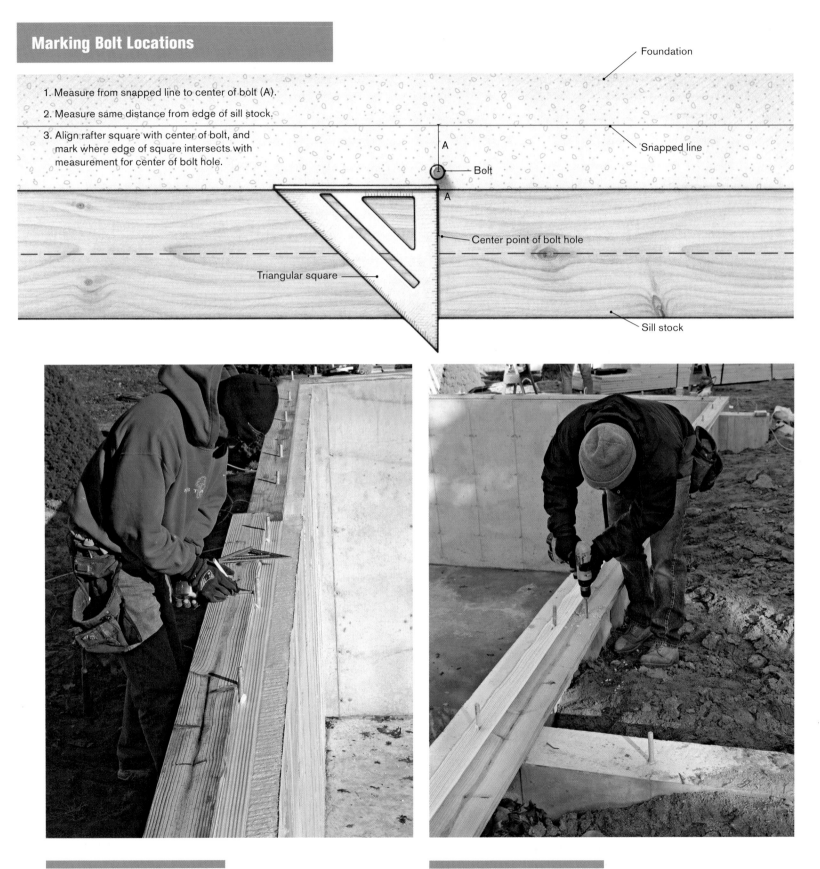

Foundation

A

Snapped line

Bolt

A

Center point of bolt hole

Triangular square

Sill stock

Set the mudsill against the bolts and measure from the chalkline to the center of the bolt. Record that distance on the mudsill.

A slightly oversize bolt hole ensures that there will be enough play to adjust the sill to the line.

Bolt-Marking Jig

A fast way to locate holes for foundation bolts is to use a homemade jig. It's simply a piece of flat metal stock 1 in. or so wide and about 1 ft. long. Cut a half circle in one end of the jig that's roughly the same diameter as the foundation bolts. The end of the jig should line up with the center of a bolt. Now measure 5½ in. (or the width of your sill stock) from the end of the jig and drill a hole big enough for a pencil. To use the jig, align the mudsill with the *inside* of the layout line chalked on the foundation. Then at each bolt, simply butt the jig against the bolt and make a mark through the pencil hole.

A homemade jig registers against the bolt and includes a hole for marking the bolt-hole location. This jig works best with 10-in. foundations or wider.

Double-Layer Mudsills

If the foundation is badly out of level or has significant high and low spots, a second mudsill can be added to the first and shimmed level before the joists are installed. You can make room for the nuts and washers by using a large-diameter drill bit and drilling partway through the stock. Holes need be only as deep as the nut and washer. Nail the second sill to the layer below, overlapping joints at the corners, and then shim the second sill to level. The second sill doesn't have to be pressure-treated wood. Wall sheathing will cover any gaps between the two mudsills.

If large square washers are used on the foundation bolts, align them with a square so they won't later interfere with the rim joists. Don't overtighten the nuts!

Once the nuts are snug, any portions of the mudsill that extend beyond the foundation should be trimmed off.

After sliding the sill stock into place over the bolts, put a washer and nut on each one. This house was built in a high-wind zone, so code required large square washers to resist wind uplift. In most areas, though, standard round washers will suffice. If you're over-zealous drilling the bolt holes, you can replace round washers with larger-diameter fender washers. Hand-tighten the nuts for now.

After all the sills are aligned with the chalklines, tighten each nut further with a wrench. If you opt to use an impact wrench, set the clutch at a fairly light tension so as not to overtighten the nuts. Too much tension can distort the mudsill, making it tough to build a flat and level floor. If using the large square washers, square them to the sill as you tighten the nuts to leave enough room for the rim joist. When the nuts are tight, go back and cut any sill stock that overhangs the foundation. As a final step, check the mudsills with diagonal measurements to make sure they are perfectly square.

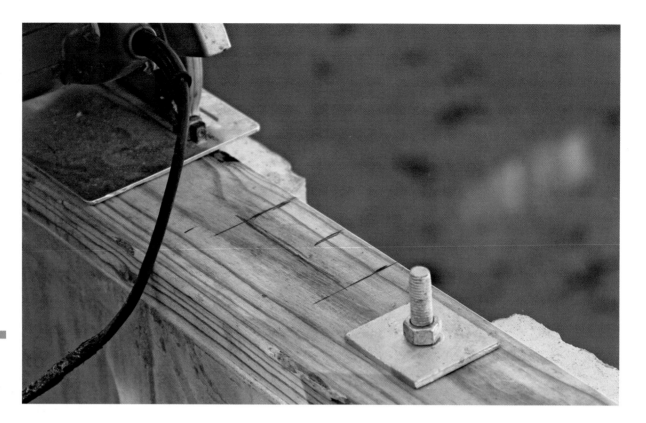

Lay out a slightly oversize notch in the plate to allow for fitting the support beam, then cut out the notch.

Support Beams

The mudsills support the ends of the floor framing, but most houses require support for the middle of the floor framing as well. Support beams or girders provide that support. A support beam can be made of various materials, including steel, LVL, and glue-laminated 2× stock. Most often, though, it is made from built-up layers of dimensional lumber such as 2×10s or 2×12s.

Locating the support beam

The support beam for this project consists of three layers of 2×12s assembled into a beam 36 ft. long. Lally columns sit on concrete pads or footings and support the beam (see "First-Floor Framing Plan" on p. 51). The beam ends typically rest in pockets formed directly in the foundation walls. These pockets are oversize to provide an air space around the sides and ends of the beam, a detail that encourages air circula-

tion and discourages rot. Beam pockets are deep enough to allow room under the end of the beam for a moisture-blocking pad.

Before you lay out the position of the beam, check the plans and confirm that the beam pockets are the right size and in the right places. You have some wiggle room but major discrepancies could require new beam pockets or a change in the length of floor joists.

If the pockets look good, lay out the location of the beam on the mudsills. Check the plans for the distance from the front of the house to the centerline of the beam and mark this point on the mudsills at both beam pockets. Now lay out the width of the beam (4½ in. in this case) by measuring from the centerline. We also marked the location for the end of the beam because the 2×6 mudsills overlapped the beam pockets and had to be notched. The layout included a ½-in. allowance for airspace at the end of the beam.

It's easy to figure out the thickness of the pad under the beam. Simply set a scrap of beam stock on the bottom of the pocket and measure up to the mudsills. A pad exactly this thickness will position the

Set a scrap of the beam material in the pocket and measure to the top of the mudsill to determine the thickness of the beam's support pad.

The crown is the natural widthwise bow of a board and must face up when the board is installed. Sight down each board and mark the direction of the crown.

top of the beam flush with the top of the mudsill. Metal plates or a piece of composite decking make a good pad that won't rot or compress.

Assembling the beam

Once the beam pockets are prepped, determine the exact length of the support beam by stretching a 100-ft. tape between the notches in the mudsills. Reduce this distance ¼ in. to ½ in. at each end so that the beam will slip in easily. Regular lumber doesn't come in 36-ft. lengths, however, so you have to build the beam out of shorter boards layered together. The joints between the boards should be staggered from layer to layer so that no two joints line up at the same place. And for maximum strength, joints should fall over a Lally column.

Start building your beam by gathering the stock. Check each board for crown by sighting down one edge. The crown is any bow across the width of the board. The crowned edge should face up when the board is installed so that the weight of the building will straighten it. Mark the direction of the crown clearly as you work through a pile of lumber. Keep the straightest boards for the longest lengths; cull the worst boards to be cut into shorter lengths.

Take the location of the support columns from the plans, and cut boards for each layer of the beam. As you cut the boards to length, make sure the ends are square. Label each board to indicate its position both by layer number and sequence in that layer.

Working from the framing plan, measure and cut individual lengths of lumber for the beam so that every joint will land over a Lally column.

Where a board in the beam bridges a joint in the layer below, drive a row of nails on both sides of the joint (note the joint in the middle layer).

A poured concrete basement floor offers a perfect working surface for beam assembly. Double-check every layout measurement as the labeled pieces go together.

On this project, the basement slab provided a nice flat surface to work on. To assemble the beam, we arranged the first layer on the basement floor. The boards for the second layer were positioned on top of the first according to layout marks made earlier, and a quick measurement of column locations confirmed the layout. The two layers were nailed together according to the nailing schedule spelled out on the plan. The location and size of nails are details that must be followed to the letter. For this beam, four 16d nails were required every 16 in. and wherever a board bridges a seam, a row of nails must be driven on both sides of the seam.

When nailing two layers of lumber together with 16d nails, angle the nails so that they don't go all the way through the other board. And as you nail the boards together, be sure to keep the layers flush at the top edge of the beam. Usually all that's needed to align the edges is a hammer tap. If more force is required, drive a toenail into the misaligned board to draw the edge flush (see "Toenailing Boards Flush" on the facing page).

When the third layer is nailed off according to the nailing schedule, carefully roll the beam over and nail through the back face as well, again nailing on both sides of any seam. As an added precaution against rot, nail a length of aluminum flashing around each end of the beam.

ESSENTIAL TECHNIQUE

Toenailing Boards Flush

There are many places in house framing where two boards must be nailed together face to face with their edges flush. Unfortunately, two boards rarely have the same amount of crown and sometimes a tap of the hammer isn't enough to make them align. In that case, drive a 16d nail at an angle through the edge of the board that must be drawn in. Then hit the nail with a hammer until the edges of the two boards line up. Drive additional nails on either side of the first if more force is needed. When the edges are flush, drive a nail into the face of the top board, near the toenails, to hold everything in place.

To align boards edge to edge, drive a toenail into the edge of the overhanging board, then pound the nail with a hammer until the board edges line up.

Setting the beam

The framing contractor on this project rented a rough-terrain forklift with a telescoping boom, a machine that makes moving lumber a breeze without straining anyone's back. More common on West Coast building sites, a lift can be rented and easily pays for itself over the course of a project. Just make sure that anyone using it has been checked out thoroughly on its safe operation.

Before a beam is lifted into place, make temporary posts to support it until the Lally columns can be set. Each post is simply a pair of 2×4s nailed face to edge to increase stiffness. To determine the height of a post, stretch a string across the sills and between the beam pockets as tight as you can make it. Then at each temporary post position, set a scrap of 2×12 (the height of the beam) on the slab on edge and measure up to the string.

Aluminum flashing nailed to the ends of the support beam provides added protection against rot.

For the height of the temporary support posts, first stretch a taut string between the beam pockets. Then measure to the string from a scrap of beam material.

A forklift or crane makes setting the heavy support beam a safe and simple operation. Slip one end of the beam into position, then the other end.

Assembling a Beam in Place

A built-up beam can be assembled in place if it's too long or too heavy to be lifted. Lay out, cut, and label all the pieces as described on pp. 62–64. Starting from one beam pocket, assemble two layers of the beam and support them with adjustable A-frame trestles. Nails staggered every few feet or so are enough to hold the layers together for now. Continue assembly to the other beam pocket, placing trestles as needed for safe, solid support.

When you've finished the first two layers of the beam, make the entire length parallel to a mudsill and hold it in position temporarily with 2×4s nailed to the plates. Then nail the two layers together according to the nailing schedule and add the last layer (or layers), checking frequently to make sure the top edges are flush, the beam is straight, and the top of the beam is in the same plane as the top of the mudsills. Many contractors use a transit to set the height of the beam, but it's just as easy to stretch a length of mason's twine between the mudsills and over the beam at every Lally column position.

When assembling a beam in place, support it on temporary trestles as you work. Keep the pieces in order and be sure to follow the nailing schedule exactly. (Note the double mudsill on this project.)

When everything is ready, roll the beam upright onto blocks and make sure the top is facing up. Next, measure and mark the exact middle of the beam. Lifting chains can then be wrapped around the beam and positioned at equal distances from the middle so that the beam is perfectly balanced as it is lifted. Once the beam clears the foundation, carefully swing it into position over the beam pockets and lower it into position. When the beam is fully seated in the pockets, set the temporary support posts in place and tack them to the underside of the beam. Be sure to set the posts where they won't interfere with the Lally columns later on.

Temporary posts nailed in a T-shape can be tacked to the underside of the beam for support.

ANOTHER WAY TO DO IT

Steel and LVL Beams

You can minimize or sometimes even eliminate the need for Lally columns in a basement by using materials that are stronger than lumber and that come in longer lengths. A steel I-beam can often span the width of a house without the need for any columns. However, steel is costly and has to be positioned with a crane or a forklift due to its weight. A wooden plate must be secured to the top of the beam with powder-actuated fasteners to provide a nailing surface for the floor joists. Also, many municipalities require that steel I-beams be enclosed in fire-code drywall before the house can be occupied.

An LVL beam is another alternative. LVLs come in long lengths and can be bolted together to form strong beams of various dimensions, and these beams are somewhat less costly than steel. The long lengths of built-up LVLs minimize joints, but the material is much heavier than conventional lumber, and a long beam may require crane placement. The density of LVL makes hand-nailing all but impossible.

A steel beam can span greater distances than wood, but you'll definitely need a crane to get it into place.

ESSENTIAL TECHNIQUE

Marking a Layout

Get in the habit of making a small V or arrow at each layout point (see the photo below). That's because a single line is tougher to read and less precise. Then, at each layout point square a line across the lumber and mark an X on the side where the framing will sit (see the photo at right). This strategy eliminates another common mistake: nailing lumber to the wrong side of the layout line.

Floor Joists

With mudsills and support beams in place, it's time to frame the first-floor deck. Take another look at the framing plan and note any special materials and details. This house, for example, has a large framed opening to accommodate the stair chase (see "First-Floor Framing Plan" on p. 51). The floor framing features dimensional lumber, but LVLs were specified to strengthen the framing in certain areas.

Floor joist layout

As mentioned in the last chapter, the most common spacing for floor, wall, and roof framing is 16 in. o.c., and that's what we used on this project. This means that the centerline of each joist, stud, or rafter is 16 in.

from the center of its nearest neighbor, unless some exception is called for on the plans. This dimension may seem arbitrary, but you'll realize its value as framing progresses: It works out perfectly to support standard 4-ft. by 8-ft. sheathing panels with a minimum of waste. Joist layout is done directly on the mudsills that run along the long dimension of the house.

To start the layout, hook the end of your tape on the end of a mudsill. It's tempting at that point to make a mark every 16 in., but that would put the *edge* of each joist on the 16-in. mark, not the centerline. This common mistake becomes painfully obvious when the first sheet of sheathing either falls short of a joist or completely covers it. Believe me, all carpenters make this mistake at least once in their career. To avoid it, subtract ¾ in. (half the thickness of the joist stock) from each of the tape's 16-in. o.c. markers. The layout marks therefore fall at 15¼ in., 31¼ in., and so forth.

To center the joists on the 16-in. layout, guidelines have to be made ¾ in. (or half the width of 2× stock) to one side of the layout points on the tape.

Mark those points, square a line across the stock, and be sure to put an X on the side of the line where the joist should fall. A corresponding joist layout has to be marked on the support beam, but in this case, we did the layout before the beam was lifted into place so the crew wouldn't have to use ladders to do it later.

The joists overlap at the support beam, so instead of marking an X on the side where the joist sits, mark an F (front joist) on one side of the line and a B (back joist) on the other side.

Once the joist layout is marked on the mudsills, go back and lay out any special framing, such as framing around the stairwell. On this house, the inside edge of the stair chase is located 12 ft. from the right end of the house, and the chase is 9 ft. wide (a measurement derived from the first-floor plan). The first step is to pull a measurement from the right end of the house to the inside edge of the chase, or 12 ft. Because there would already be a joist at that location, we opted to put the LVLs right next to the joist, which moved the LVLs in a bit (¾ in.) but not enough to affect future framing. From there we measured the overall width of the chase (9 ft.) to locate the LVLs for the other side.

The last step in joist layout is to snap a chalkline to indicate where the ends of the joists will fall. This line also locates the inside face of the rim joist.

Joists made of dimensional lumber overlap at the support beam, so an F is marked on one side of the line to indicate where the front joists should land; a B locates the back joists.

A straight line snapped on the mudsill guides the placement of the joists while providing space for the rim joist. (Local wind-uplift codes required the anchor bolt details shown here.)

Measuring Joist Spacing

An alternative way to lay out the joist positions is to mark the edge of the first joist for the proper on-center alignment. Partially drive a nail at that mark and hook your tape on the nail's head. Then mark the rest of the layout at the 16-in. increments shown on the tape. I'd use this method only for the 16-in.-o.c. layout, however. Measurements for any variations in the framing should always be taken from the edge of the mudsill to ensure accuracy.

Make the first mark at 15¼ in. Then hook your tape on a nail and make layout marks on the 16-in. increments.

Installing the joists

On this house, the joists were installed in two stages. Joists on the front half of the house were installed first, followed by those on the back half of the house.

The location of the support beam meant that 12-ft. joists would be sufficient at the front of the house. As each joist was pulled from the pile, it was checked to make sure the sill end was square. Then each joist was laid flat over the support beam and the mudsill in a process called *loading*. Loading joists takes a bit of practice but it's simply a matter of sliding one joist over another that's already in place. This makes loading the joists a safe one-man job. As joists were loaded, a crew member crowned each one and drew an arrow to indicate which edge should face up. To avoid errors later, the crown arrows were oriented so they all faced in the same direction as the joists lay flat.

The first joists installed are meant to keep the support beam straight. If you merely set joists from one end of the house and work toward the other end,

you can just about guarantee that the beam will get pushed out of position. So in this case we chose a joist on each side of the stair layout as the first joists. At the right end of the house, we measured from the rim joist layout line to the edge of the support beam, and rolled the first joist up on edge. Then we transferred this measurement to the bottom edge of the joist and made a layout mark. The crew member at the mudsill toenailed the joist to the mudsill, aligning it with the rim joist chalkline. A second crew member tapped the beam until it aligned with the layout mark on the joist and then nailed the joist to the beam. We repeated that procedure until four joists held the beam straight over its entire length. After that, it was safe to roll and nail the remaining joists, starting with the rim joist at one end. When all the joists on the front of the house were secure, we set the LVLs for the stair chase. LVLs come a full 10 in. wide, so each one had to be ripped down to 9¼ in. to match the actual width of a 2×10. After

To put the joists in place, slide them over to the beam on top of its neighbor already lying flat between the mudsill and support beam.

When the joists are loaded and ready to install, crown each one, mark the direction of the crown with a V, and point all the Vs in the same direction.

The first joists nailed into place hold the beam straight and prevent it from slipping out of position as subsequent joists are installed.

When the beam has been stabilized, roll the remaining joists on edge and nail them to the layout.

sliding the LVLs into place on their layout marks, they were nailed into place.

Once the joists at the front of the house were in place, we tackled the back joists. To load the back, joists were piled on the forklift so that the crew could lay them out easily (see the top right photo on p. 72). These joists had to be just over 14 ft. long to overlap the beam, which meant that 16-ft. lumber was specified on the materials list. But rather than waste the extra couple of feet by letting it overlap the front joists, we cut the excess off each joist and stacked it to use for blocking later.

Layout and installation of the back joists were straightforward, except over the bulkhead opening in the foundation. Here the rim joist had to be doubled to create a header that would span this opening, so the joists that landed over the bulkhead were moved inward 1 ½ in. to accommodate an extra thickness of

LVL beams that beef up the stair opening slide into place next.

With the beam held in place by the front joists, the back joists can easily be loaded and installed.

rim joist stock. This extra length of rim joist was cut to fit between the joists closest to the sides of the bulkhead and was nailed to those joists (see the photo below). When the rim joist was later installed, it formed the second half of the header.

The bulkhead required a header to span the opening, so an extra layer of rim joist stock was nailed to the shortened joists. The main rim joist itself will form the second half of the header when the two pieces are nailed together.

Installing rim joists The rim joists that run parallel to the floor joists are installed with the joists. The perpendicular rim joists—the ones that cap the ends of the joists—can either be installed as the floor joists go in or be added after the joists are in place. Their role is to hold the ends of the floor joists plumb and square to the mudsills.

Start by nailing the ends of the parallel rim joists into the perpendicular rims at the corners. Next, toenail the rims into the mudsills to draw them tight against the bottom corners of the joists. Now hook your tape on the end of the rim and line up the center of the first regular joist with the layout mark on the tape. Drive a nail through the rim and into the top of that joist to keep it in position. Hook your tape on the edge of the first joist and line the next joist on the layout mark. Secure it with a nail at the top. Keep the tape hooked on that first joist and continue down the row, nailing the tops of all the joists on the layout. Check as you go to make sure the joists are square to the top edge of the rim joist and not slightly twisted. When the top of each joist is secured with a nail, go back and drive the rest of the nails through the rims into the joists.

Installing headers The last major piece of floor framing to be installed on this job was the LVL header that fit between the doubled LVLs at each side of the stair chase. After the location of the header was laid out, we nailed temporary blocks to the bottom of the LVLs to support the header. Once the header was nailed into place, we removed the blocks and slipped metal joist hangers into place. Short stub joists completed the framing from the header to the support beam. Joist hangers were added later to connect the stubs to the LVL header.

The last joists are short lengths that run from the support beam to the stair-chase header.

Floor Sheathing

When floor sheathing is glued and nailed to the joists, the result is a strong, flat, and stiff structural system. Sheathing also provides a base for the wall framing and the finish flooring. In most cases, carpet and wood strip flooring can be installed directly over the sheathing without additional underlayment.

Deck sheathing is typically plywood or, as it was in this case, ¾-in.-thick tongue-and-groove (T&G) OSB. The T&G edges are on the long sides of the panels so that each course interlocks with the next. The ends of the panels do not have T&G edges, so they must always fall on a joist or on solid lumber blocking.

Running the first course

Take quick diagonal measurements across the joists to make sure the framing assembly is still square. Where this house was built, the building code required that solid blocking be installed in the two outermost joist bays every 4 ft. (another wind-related requirement). Before sheathing began, the crew made layout marks every 4 ft. for the blocking and installed it.

It doesn't usually matter where sheathing starts, but in this case we opted to start at the back of the house. A chalkline snapped across the floor system 4 ft. from the outer edge of the rim joists served as a guide for the first course of sheathing (see the top left photo on p. 74). To install sheathing, run a bead of construction adhesive on the top of each joist, stopping just short of the chalkline. (Don't put adhesive on

Mark the layout every 4 ft. to identify spacing for the block-ing and sheathing. A line snapped at the first mark guides the first course of sheathing.

Construction adhesive prevents squeaks and stiffens the floor system. A sheet of floor sheathing provides a safe and comfortable platform from which to work.

The first piece of sheathing should be placed carefully to avoid smearing the adhesive. It's a job for two people.

the rim joists because you may have to adjust their position later.) If you spread the adhesive beyond the layout line, it can be a slip hazard (and it gums up work boots something fierce). If you have a large enough crew, one person can spread adhesive as others place sheathing. Otherwise, spread adhesive for one sheet at a time to prevent it from drying out. Remem-ber that a double bead of glue is needed where sheets of sheathing meet at a butt joint.

Carefully drop the first full sheet in place with the tongue facing the rim joist (subsequent courses will be

driven into place with a sledge, which would damage the tongue). Don't just flip a panel haphazardly onto the joists and slide it into position, because this smears the glue and makes it useless. Instead, two people should drop the sheet gently onto the joists as close to final position as possible. Move the sheet slightly to the snapped line and align it to the center of the last joist it lands on. Tack the sheet in place (see the top photo on the facing page) and double-check its alignment before nailing it. Drive nails at the ends of each joist but don't nail into the rim joist just yet (this happens later). To make sure the joists stay straight, measure from the end of the sheet and nail the first joist exactly on its layout spacing. Now you can hook the tape on the secured joist and nail the rest of the joists at their proper spacing. Many manufacturers now print a centering mark every 16 in. on floor sheathing to make it easier to keep the joists in line.

Install the rest of the sheets in the first row in the same manner. If necessary, cut the last sheet to fit before installing it. This house was 36 ft. long, so a half sheet worked perfectly to complete the courses on the back of the house. Nail off the first course completely (except for the rim joists) before moving to the next course. That course has to be securely in place before the next course can be driven against it.

Temporary Work Platform

It's tempting to balance on open joists when installing sheathing, but it's not a good habit to get into. Any joist can wobble unexpectedly and toss you into the basement. Whenever possible, lay sheathing over the joists to serve as a temporary work platform, as shown in the top right photo on the facing page. Just make sure the ends of the sheet are fully supported.

Completing the sheathing

Butt joints in successive courses of sheathing should be staggered so that they don't line up, and in practice this means offsetting them by a half sheet. So we started the second course with the 4-ft. length of sheathing left over from the end of the first course. Whenever possible, use the factory edge of a cut sheet to butt against the adjacent sheet to provide the tightest fit.

Snap a chalkline for the second course to indicate where the adhesive should stop, then run a bead of adhesive on the tops of the joists and squeeze a bead into the groove of the mating sheet on the first course. Experienced framers have an interesting technique for placing sheets for successive courses. They hold a sheet

When the first sheet is in place, tack a corner to hold it at the centerline of the joist (top). Then nail it to the remaining joists, guided by the layout marks printed on the sheathing (above). This prevents the joists from wandering off course as subsequent sheets are installed.

The rest of the first row of sheathing goes on in the same way as the first sheet. Nail the first course completely so the sheets will bond to the adhesive and withstand the pounding needed to get the subsequent rows into place.

Begin the second row with a half sheet so that the seams of subsequent sheets won't line up with those in the first row. As the sheet drops, keep a foot on it to keep it from sliding out of position.

To mate the tongue and groove edges, have a helper stand on the edges as you drive the sheet into place. When working solo, hold the edges in line with one foot as you tap with the sledge hammer.

vertically to start, resting it on the tongue edge. Next they lower the sheet with their arms extended and with one foot up on the lower portion. When they finally let go of the sheet, they maintain pressure with their foot to keep the sheet from bouncing out of position or sliding forward. This process, more ballet than carpentry, effectively gets the sheet ready to be driven into place without smearing the adhesive.

When the sheet is down, drive it gently into place with the sledge hammer until the tongue mates with the groove of the previous course. A 2× scrap keeps the sledge from damaging the groove. Quite often the sheets don't sit perfectly flat and the tongue can't find the groove without help. By standing on the tongue edge, one person can align the edges while the sheet is being driven. If you're working alone, stretch one foot to the mating edge while tapping with the sledge.

If the sheet bounces out of alignment as you nudge it with the sledge, drive one corner into position and tack it with a nail, then tap along the edge of the sheet until the rest of the tongue slips into place. Pay attention because tapping one end of a sheet can make the other end separate from its neighbor. Before nailing

off the sheet, make sure each end seam is tight. If it isn't, just tap the sheet gently until the gap disappears. Continue adding sheets and courses until you reach the other side of the house. By working one complete course before starting others, you can correct small misalignments before they snowball into big ones.

Once the back half of the house is fully sheathed, start on the front half. However, remember that the floor joists don't usually run in continuous lengths from the front of the house to the back: They overlap at the support beam (see the top photo on p. 75). That calls for a slight change of sheet layout. At the start of the first course for the front of the house, simply let the end of the sheet overhang the rim joist by 1½ in. so that the other end will fall properly on the centerline of a joist. You can trim off the overhang later. The alternative? You'd have to nail 2× scrap stock to every joist in order to support the ends of each sheet in that course, a time-consuming and wasteful exercise.

At openings in the floor system (such as for stairs), you can let the sheathing overhang a corner, nail it in place, and cut it to fit later, or you can cut it to fit before nailing it. If you cut first, be sure to resume the

Preventing Accidental Falls

At this project, the stair chase left a very large opening in the floor system. We opted to sheathe over it entirely so nobody would accidentally fall through it as framing continued. To support the sheathing we added a couple of temporary joists to span the opening.

course on the other side of the opening in the same pattern and layout as the rest of the floor system.

Most likely you'll need to rip sheets for the final course of sheathing. If the rip is less than the width of the wall framing that will fall on it (less than 5½ in. for 2×6 walls), it doesn't need a tongue or groove.

Adjusting the rim joists The joist ends are square, so the rims nailed to the ends of the joists have to be plumb and straight, right? Not necessarily. All the pounding during the process of nailing off the joists and sheathing the floor can move the rims slightly, and there are always slight variations in the ends of the joists themselves. If you want the wall sheathing to fit properly later on, the rims have to be in the same plane as the first-floor wall studs. That means they have to be straightened and plumbed before you nail the edge of the sheathing.

Start the process by sighting the rim joist. If it's not perfectly straight, plumb up with your rafter square from the framing at each end of the rim, then measure in a couple of inches and snap a chalkline on the sheathing. Now at every third or fourth joist location, place the edge of your square against the rim joist and measure to the line. If the rim needs to move out, pry it with a flat bar. If it needs to move in, bang it with your framing hammer. When the measurement is right, nail

the sheathing into the rim at that point (there's no need to squirt construction adhesive between the two). Work your way down the entire rim in that manner before going back and nailing off the whole rim. If you find a large section where the rim needs to be moved a lot (⅛ in. to ¼ in.), you may have to straighten and nail the rim at each joist location.

On this project, blocking in the outer joist bays kept the rims fairly straight, but sheathing hung over the rims in a few places. So we snapped a line to guide a circular saw and then cut back the sheathing before straightening and nailing off the rims.

To straighten the rim joist, snap a straight line and measure over to the line from the face of the rim. Adjust the rim in or out as necessary, and then nail it to the edge of the sheathing.

Installing Lally Columns

As soon as the first-floor deck is complete, Lally columns must be installed to replace the temporary posts under the support beam. The first step in the process is called *blocking and stringing* the beam. Nail 2× blocks to the bottom of the beam at each end, then drive a nail partway into each block and bend it toward the outside wall. Hook the end of your chalkline to one nail, then pull it *really* tight to the other nail and secure

it there. Now hold a 2× block against the string at each column location. If the beam is level, the block should just barely slip under the string. Chances are, though, that the weight of the floor system has caused the beam to sag slightly. Now's the time to fix that.

At each column location, nail a Lally column plate to the underside of the beam. Drop a nail from the center of the plate and watch where it lands to determine the approximate column location on the floor. Put a plate on the floor at this point. Start at the first column location and raise the beam slowly, using a hydraulic jack and a 4×4. When the string and block show that the beam is at the proper height, measure the distance between the plates, transfer this measurement to a length of Lally column and cut it to length.

To install the column, jack up the beam an extra ¼ in. or so, place the column carefully onto its floor plate, and then slip the top into place. Lower the beam until the top plate captures the top of the column, then plumb the column front to back and side to side. Finally, release the jack and remove the temporary post.

Setting columns gets more complicated after a couple are in place because jacking the beam can loosen adjacent columns, which can then fall over. Always have a crew member hold a nearby column in place until its neighbor has been safely set in place.

Lally column plates have little nubs that hold the column in place once it is under load. In addition, some locales may require that the plates be spot welded to the tops of the columns. If the bottom plates sit on top of the finished basement floor (as with this project), they must be secured with masonry nails.

The LVL beams around the stair opening had to carry a lot of weight in this house. When the crew finished installing the columns, we opted to frame support walls under those beams.

To gauge where the support beam might be sagging, stretch a line over blocks tacked to each end of the beam (top left). Another block held against the string shows where, and by how much, the beam must be raised (bottom left).

Cutting a Lally Column

The tool for cutting a Lally column is a giant version of a plumber's pipe cutter. Align the cutting wheel with the length mark on the column and tighten the cutter around the pipe. As you rotate the cutter around the column, make sure the cutter stays in the same groove. If the cutter is misaligned, it will travel in a spiral and won't cut the column properly. While one person spins the cutter, another has to keep the column from rotating; clamps or a large pipe wrench can help with this task. After every couple of turns, tighten the cutter slightly. When the cutter finally breaks through the metal exterior of the column, the waste end falls away (be sure to keep your feet at a safe distance!). Inspect the end of the column and chip off any concrete that sticks out past the metal casing. If you don't remove it, the column won't fit properly.

Nail a metal plate to the beam at each column location, then position the bottom plate. With the beam jacked to the proper height, measure between the plates to determine the length of the column.

Drop the beam back down enough to capture the column securely, and then tap it with a sledge hammer to plumb the column in two directions.

Safe Jacking Posts

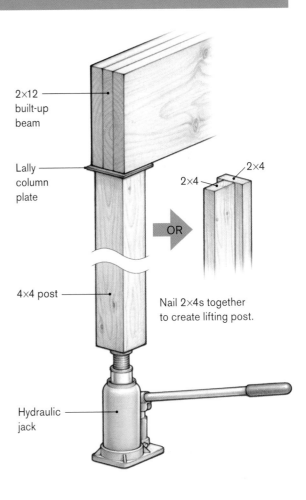

2×12 built-up beam

Lally column plate

4×4 post

Hydraulic jack

2×4

2×4

OR

Nail 2×4s together to create lifting post.

4

Exterior Walls

In the last chapter, I emphasized the importance of keeping things square and level, but in this chapter *plumb* is the byword. Just as installing an out-of-square deck creates problems that affect subsequent framing, an out-of-plumb wall will haunt you forever. You may have to measure a deck to realize that it isn't square, but the human eye seems to zero in on a wall that's out of plumb just as it notices paintings hanging crooked.

I've heard professional framers argue passionately about the best way to build a wall. Is it better to sheathe the walls before they're raised or after? Should the window openings be cut when the walls are flat or after they're up? Heck, I've worked with framers who built walls upright and toenailed all the studs to the plates. The method I'll describe in this chapter is pretty close to the one I first learned, but I'll point out alternative techniques where I think they make sense.

Layout and Preparation for the First Walls

The plans for our house called for studs to be spaced 16 in. o.c. In some regions 2×6 walls can be built 24 in. o.c., which saves a lot in materials. From a construction standpoint, however, the only difference between the two is a matter of spacing, not technique.

Many houses are built with 2×4 studs, resulting in stud bays that are 3½ in. deep. This depth easily accommodates fiberglass batt insulation rated R-11 (a measure of its ability to resist heat flow), which is just fine for some climates. But the house shown here is in New England, so we used 2×6 studs and plates, which gave us 5½-in.-deep bays. The extra depth means more space for insulation: R-19 or better. However, the techniques used for framing 2×4 and 2×6 walls are essentially the same.

STEP BY STEP

How to Frame an Exterior Wall

Whether it's made from 2×4s or 2×6s, each exterior wall is built in pretty much the same sequence:

1   Cut plates to length.
2   Mark the framing layout on the plate edges.
3   Assemble corners and partitions.
4   Install corners, partition assemblies, and window and door framing.
5   Install the studs.
6   Sheathe the wall.
7   Tilt the wall upright and brace it.
8   Nail the bottom plate to the floor system.
9   Plumb and brace the wall.

Snap guides for the walls

Use your square to plumb up from the rim joist on the front of the house and then measure 5½ in. from the square. (You can't just measure in 5½ in. from the edge of the deck sheathing because the edge might not be even with the face of the rim joist.) Mark a point 5½ in. in from the rim joists at each end of the front

Guide lines snapped on the deck aid in positioning the plates. Take your measurements from the rim joist, not the sheathing edge, and double-check the lines after snapping them.

wall and snap a line between the points. Do the same for the back wall. These lines correspond to the inside face of the wall framing and ensure that the walls will be straight.

Mark points on the line for each wall to indicate where the inside edges of the end walls meet it.

After the plates are cut, toenail the bottom plate to the deck along the snapped line (top), and then tack the top plate to the bottom plate so they can be laid out at the same time (above).

Measure between the corner points to double-check that both the front and the back walls are exactly the same length. If the wall measurements are not the same, maybe you marked the wrong measurement or didn't plumb up from the rim joist properly. Go through the process again. If the measurements are still off, relocate the end points of the walls slightly to compensate for the difference. Be sure to split the discrepancy between both walls. In other words, if there is ½-in. difference in the wall measurements, add ¼ in. to the shorter wall and subtract ¼ in. from the longer wall to make the adjustment less noticeable. Now take diagonal measurements between the inside corner points to make sure the layout is square.

Cut the top and bottom plates

It's wise to build the front and back walls of the house first because they're longer and heavier than the end walls. Besides, you'll need plenty of elbow room to wrestle them into place. The first step in building the walls is to cut the plates.

You probably won't be able to use a single piece of stock for each plate. No problem: Make them in sections. We divided each 36-ft.-long wall into three sections, and cut a top plate and bottom plate for each one. The length of each section isn't important, and the sections don't all have to be the same length, so work with the lengths of stock that you have available. The important thing is that each pair of plates be identical in length. Once we cut the plates for the first two sections of each wall, we placed them in pairs end to end along the chalked layout line and measured to determine the length of the remaining section. When the plates are complete for one entire wall, toenail the bottom plate to the layout line every 4 ft. or so using 16d common nails. The nails hold the plate straight during layout and act as hinges later when the wall is raised.

The easiest and most consistent way to lay out the plates is to mark the top and bottom plates at the same time, so tack the top plate to the bottom plate every 8 ft. or so. Where the plate from one section meets the plate for the next, toenail the ends together through the edges to keep them as one during layout.

First-Floor Plan: Exterior Walls

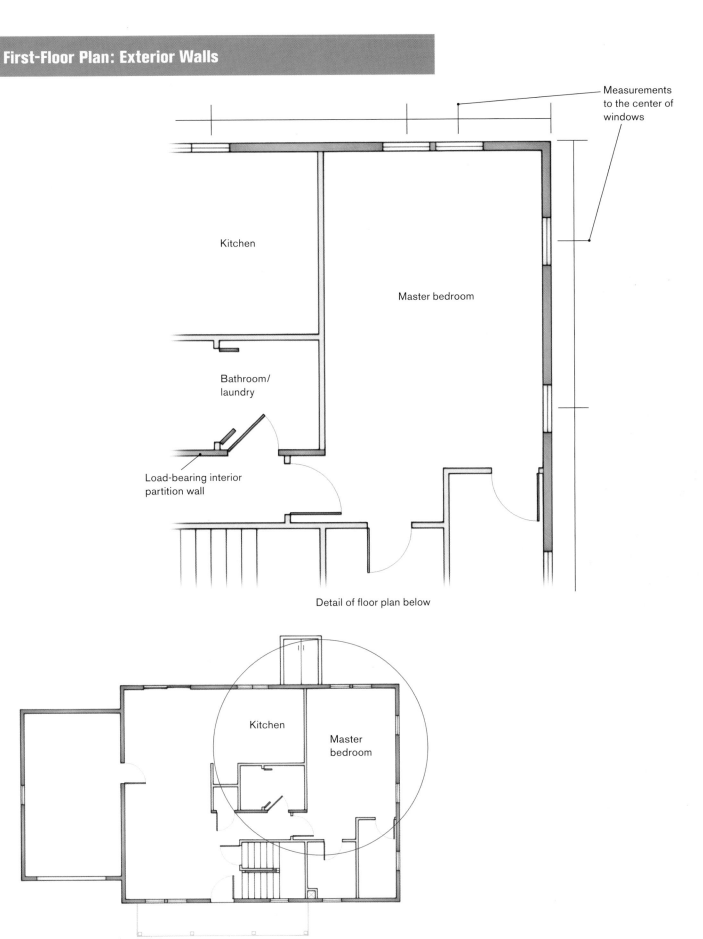

Measurements to the center of windows

Kitchen

Master bedroom

Bathroom/laundry

Load-bearing interior partition wall

Detail of floor plan below

Kitchen

Master bedroom

Lay out the openings

The floor plan shows measurements to the centers of the window and door openings, measured from one end of the wall plates (see "First-Floor Plan: Exterior Walls" on p. 83), so those center points should be the first to be located on the plates. Following the plan, measure from the center of one window or door opening to the center of the next and mark these points as a centerline. A centerline mark is the letter C with an L going through it (see the symbol at left). At each window location, mark half the width of the rough opening (r.o.) to each side of the centerline. These marks represent the face of the innermost jacks and set the width of the rough opening. Then mark and label the location of the king studs and any other jack studs on each side of the opening (see the top photo on p. 86). The number of jacks and kings needed for a window or door opening can vary. An unusually large window, for example, might need an extra jack on each side to support the long header. In this house we had to *triple* the kings and *double* the jacks to meet the local code (see "Framing for a Window Opening" on the facing page). This requirement was specifically noted on the plans by the engineer.

Squaring the Lines

Layout is just a series of marks and symbols on the edges of the plates, but each mark must eventually be squared across the wood using a triangular square. The squared lines make it easier to align the studs and jacks later. Some framers square the layout marks across the plates as they do their layout, while others wait until all the layout marks are made and then go back to square the lines. It's just a matter of personal choice.

Always measure for your layouts by pulling the tape from the same side of the house. A consistent layout keeps the framing members aligned as you work and helps minimize mistakes.

The centerlines of the window and door openings are the first points marked on the plates. Measure from those points and mark the sides of the rough openings next.

Framing for a Window Opening

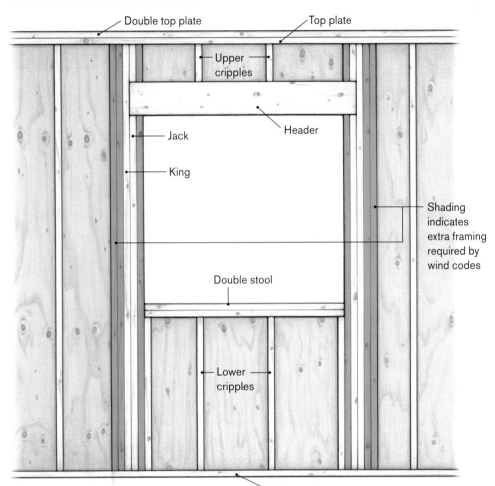

Double top plate

Top plate

Upper cripples

Header

Jack

King

Shading indicates extra framing required by wind codes

Double stool

Lower cripples

Bottom plate

Rough Openings

Window and door manufacturers specify the size of the rough opening for each of their products. This number is the actual outside dimension of the window or door plus a bit extra to allow for fitting. Every framer I've ever worked with took great pains to build rough openings square, but when a wall is raised that's hard to maintain. There always seem to be small discrepancies, hence the term *rough opening*. The extra space in r.o. dimensions is there to accommodate such discrepancies.

Beefed-Up Openings

Because this house was built in a high-wind area, the engineer specified extra framing for each opening. Typical framing for an opening in an area with no extra code restrictions would consist of a single king and a single jack stud for each opening. For the openings in this house, the king studs were tripled and the jacks, doubled.

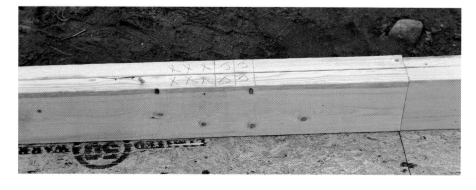

On both sides of the rough opening, mark the position of king studs and jack studs. Plans for this house called for triple kings and double jacks at each opening. Plate sections should be edge nailed together to prevent them from separating during layout.

Mark the position of studs and cripples after the layout for the openings is complete. Always pull your tape from the same end of the house to ensure accurate dimensions.

Lay out other wall details

Once the layout for the rough openings is complete, indicate where interior walls meet the exterior walls, taking the measurements from the floor plans. Mark the actual width of the partition framing on the plates at these locations, which is also width of the nailer in the partition backer. Then draw the position of full-length studs on either side of the nailer to complete the layout. But be sure to note on the plate which side of the wall the nailer should be on. Believe me, it's embarrassing to raise the wall and discover that the nailer is on the wrong side!

Finally, mark the layout for studs and cripple studs. If the wall is longer than your tape measure, partially drive a nail at a layout point and hook the end of your tape over it to continue the layout (such a nail is visible in the top photo above).

Preassemble the Walls

Once you finish the plate layout, separate the top plate from the bottom plate and move it about 8 ft. away in preparation for wall assembly. But leave the bottom plate toenailed to the chalkline (you'll see why later). Position the studs between the plates, but don't nail them in until other parts of the wall are assembled or they'll just get in the way. The studs for the project house were used as delivered, but all other parts of the wall were cut to length on site, including the jack studs, headers, stools, and cripples.

Studs

Framing stock typically comes in lengths of even feet: 8 ft., 10 ft., 12 ft., and so on. For this house, 8-ft. studs were specified for all the wall framing. However, many houses are framed with precut studs (see "Precut Studs" on the facing page), which are actually a little shorter than 8 ft. Using full-length studs results in slightly higher ceilings, but standard 4-ft. by 8-ft. sheets of drywall won't quite cover the taller wall. No problem: The drywall guys added narrow strips to the bottom of the wall to make up the difference, and the trim carpenters installed tall baseboard to hide the extra drywall seam.

Other parts of a wall

Many framing elements are made from multiple pieces that are easier to assemble before they are nailed to the plates. That includes corners, partition backers, and jack/king assemblies for the sides of the openings. As each one is complete, you can place it in position between the plates.

Corner assemblies There are many different ways to build corner assemblies (see the drawings on the facing page). The most common configuration is a simple L-shape, also known as a two-stud corner: one stud nailed to the edge of another. However, the engineering for this house called for each wall to end

Three-Stud Corner

Advantages: Quick to assemble, strong, good support for corner boards and siding

Disadvantages: Not ideal for 2×6 framing, uninsulated corner

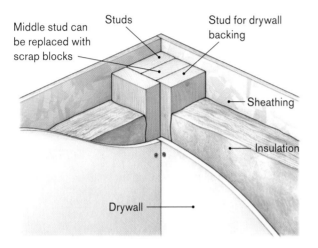

Middle stud can be replaced with scrap blocks

Studs

Stud for drywall backing

Sheathing

Insulation

Drywall

Two-Stud Corner

Advantages: Uses less lumber, easy to insulate

Disadvantages: Less support for corner boards and siding

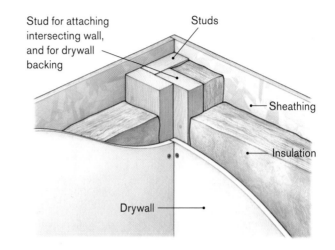

Stud for attaching intersecting wall, and for drywall backing

Studs

Sheathing

Insulation

Drywall

ANOTHER WAY TO DO IT

Precut Studs

If the house design calls for 8-ft. ceilings, then a little has to be cut off every 8-ft. stud to put the ceiling at the right height. Those little cut-offs aren't much use beyond kindling for the general contractor's fireplace. So to save material, lumber companies in my part of the country sell studs precut to a length of 92⅝ in.

Here's where that figure comes from. To end up with an 8-ft. (96-in.) finished ceiling height, the framed wall actually has to be 97⅛ in. from the subfloor to the ceiling joists. The extra 1⅛ in. accounts for the thickness of ⅝-in. ceiling drywall and ½-in. finished flooring. If the studs have to be 4½ in. shorter than the total height of the wall (the thickness of two top plates and one bottom plate), then that leaves us with 92⅝-in. studs, for a finished ceiling height of 8 ft. The general contractor can find kindling elsewhere.

Boxed Corner

Advantages: Good for 2×6 framing, can be insulated

Disadvantages: Insulation must be added before sheathing is installed

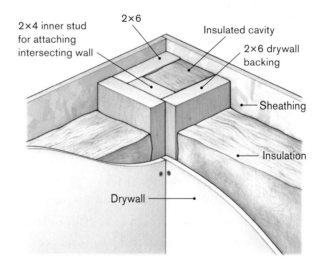

2×4 inner stud for attaching intersecting wall

2×6

Insulated cavity

2×6 drywall backing

Sheathing

Insulation

Drywall

The corner assemblies for this house are L-shaped, with the end studs doubled. The other leg of the L provides a nailer for the intersecting outside wall. In the background, a completed corner assembly is ready for installation.

To assemble the sides of the openings, nail the jacks to the first king stud. Add the other kings after the header has been installed.

A header for a 2×6 wall consists of three 2×8s sandwiched around two lengths of ½-in. plywood.

with a double stud. So we nailed the L together first, then added the second stud, as shown in the top left photo above.

Jacks and kings
Jacks sit beside the king studs and support the header. The length of the jacks is usually determined by the height of the door rough opening, which can be found on the plans. Cutting the window jacks the same length as the door jacks makes the tops of the windows line up with the tops of the doors, which is normally a desirable design element, but be sure to check the plans for any variations.

Nail the jacks to the king studs for each opening. For this house, each opening had two additional king studs on each side, but we didn't nail them into place until the header was installed so that we could nail through the first king stud and into the end of the header.

Headers
A header spans the opening between the innermost king studs, so its length can be taken directly from the layout on the plates. In some areas of the country, headers are made from solid lumber, but the 6×8s needed for this project would have been pretty expensive. A more economical way to build a header for a 2×6 wall is with three 2×8s assembled on edge, separated by two spacers of ½-in.-thick plywood. You can rip a bunch of spacer strips to width before assembly begins and then cut them to length as needed. Each plywood strip should be ½ in. less in width and length than the size of the header so it won't stick out beyond the edges of the header.

To assemble a header, sandwich a length of plywood between two 2×8s. Line up the 2×8s using a triangular square and tack the sandwich together at the four corners. Now add the second piece of plywood and place the third 2×8 on top, aligning it with the pieces below. Tack the corners again and then

Partition backers provide nailing for intersecting 2×4 interior walls.

drive four 16d nails every 16 in. across the face of the header. Flip the header over and drive the same nail arrangement from the other side.

Partition backers In this house, all the interior walls were framed using 2×4 construction, so each partition backer consisted of a 2×4 with 2×6 studs nailed to each edge. The studs on the edge act as drywall nailers.

Assemble the Walls

Once all the framing pieces are in position, you can nail them to the plates. Come to think of it, it's the other way around—you nail through each plate and into the end of each vertical framing element. This type of nailing results in a stronger wall, and it's a lot easier than trying to toenail studs to the plate!

It's important to make the edges of the framing flush with the edges of the plates so the sheathing and interior wall finish will fit properly. That concept might seem easy enough, but first-time framers usually struggle with this task because it's tough to keep the parts aligned while driving nails. Don't try to hold a stud in place with your hand because it might get punctured by a pneumatic nailer or clobbered by an errant hammer swing. The time-honored solution is to use your foot instead (see "Nailing Studs into Place" on p. 90).

Assemble the openings

Many framers start at one end of a wall and work to the other end, nailing all the pieces as they go. But I prefer to nail the king/jack assemblies to the bottom plate first and then nail the other elements. This way the regular studs don't get in the way when you have to nail through the sides of the framing to assemble the openings. If the studs are already in place they'll limit your access. After nailing the jack and king stud assemblies to the bottom plate, I put a header in place and nail through the king studs into each end of it. In this project, we nailed the extra king studs to the assembly at this point to meet local code requirements.

Door rough openings are complete once the header and jacks are installed. The bottom plate will be removed much later to make room for the door.

Once the header is in place, nail into it through the king studs on each end. Then add any extra kings if required by code.

Windows are more complex because stools and cripples must be cut and installed to frame the bottom of each opening. The lengths of cripples can be determined by checking the window schedule on the plans. First subtract the rough-opening height from the length of the jack, then subtract 3 in. more to account for the thickness of the double stool (if you're using a single stool instead, subtract 1 ½ in.). The number remaining is the length of the cripples. The cripples above a window, if any, are cut from scrap stock later on and aren't usually cut ahead of time. Just don't forget to put them in!

Regular wall studs don't interfere with framing the bottom of a window opening, so they can be installed at this point. To complete the window opening, first nail in the cripples that go alongside the jacks, as shown in the top photo on the facing page. Then nail through the bottom plate and into the other cripples, using the stud layout marks as your guide. Leave the tops of the cripples loose for now.

Cut a stool to fit between the jacks and tap it into place over the ends of the cripples. Then hook your tape on a nearby stud on a regular layout and transfer the stud layout to the stool. Nail through the stool and

ESSENTIAL TECHNIQUE

Nailing Studs into Place

To keep the studs in alignment as you nail into them through the plate, keep your heel on the stud and your toes on the plate. Your weight on the joint pins both parts to the subfloor, ensuring that the edges are flush. Tap the stud with your hammer as you step on the joint to align it with the layout mark, and hold it in place with your foot as you drive the nail.

Use your foot to align a full-length (8-ft.) stud with the plate as you nail. Pressure on the stud keeps the top edges lined up, and you can fine-tune the position of the stud with a hammer tap or simply by twisting your foot.

Nail cripple studs to the jacks at the bottom of the window rough opening. Here one cripple is already in place as a crew member nails in the second.

into the ends of the cripples. Another way to mark the cripple layout on the stool is to set the stool against the bottom plate and transfer the layout directly.

Once the stool is secure, nail the second stool to the first. Second stool? I've heard builders argue that a second stool is a waste of time and material because it isn't necessary from a structural standpoint and isn't required by code. However, as someone who has done a lot of interior finish work, I can tell you that the extra stool makes it a lot easier to nail on the apron (the bottom part of window trim) later on. To me, that makes the extra stool a no-brainer.

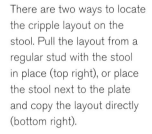

There are two ways to locate the cripple layout on the stool. Pull the layout from a regular stud with the stool in place (top right), or place the stool next to the plate and copy the layout directly (bottom right).

Mulled Windows

The design of this house called for a number of single windows installed close to each other in pairs. One way to handle this is to order the windows with a mullion, a piece of trim that joins them into a single unit. But as a design element in this house, individual windows were separated by more than the width of a factory-built mullion. We decided that the easiest way to frame this detail was to build a single opening for both windows, with a post in the middle.

If there are any cripples above the header, install them now. Because these pieces are usually fairly short, cut them out of scrap if you can do it safely. Nail into each one through the top plate and toenail the other end into the top of the header.

Install studs, corners, and partition backers

Now you can assemble the other wall components. At this point, it makes sense to start at one end of the wall and work your way to the other. As with the opening assemblies, nail the pieces to the bottom plate first.

At the corners, drive three 16d nails into each leg of the L. And make sure the flat part of the L (the nailer for the end walls) is facing down against the deck. Partition backers get three 16d nails in each 2×6 stud and three in the 2×4 nailer. Regular 2×6 studs also get three 16d nails in each end. When all the components are nailed to the bottom plate, nail off the top plate by working your way from one end to the other.

Install the second top plate

When all the wall framing is complete between the plates, double-check it to make sure everything is secure—it's easy to forget a nail here and there or to skip nailing a stud. At this point, install the second top plate. It should always be installed before long walls are raised because it bridges the butt joints in the first top plate and strengthens the wall. To increase the stiffness of the wall, install the second top plate in lengths that bridge any joints in the first top plate by a couple of stud bays. It's crucial that the edges of the two plates stay aligned.

The second top plate isn't the same length as the first one, as you can see in the top photo on the facing page. Cut it shorter at each end by 5 ⅝ in. That measurement leaves a space that allows the second plate of the intersecting wall to lap over the corner at this point, tying the walls together. Why 5 ⅝ in. and not 5 ½ in., the width of a 2×6? A little extra space makes it easier to fit the intersecting plate.

Square the wall

As a final step before sheathing the wall, make sure the wall is square by taking diagonal measurements with a 100-ft. tape. Use a sledge hammer to tap one corner as needed until the measurements match. Once the wall is square, tack one corner to the deck to keep it in position until the sheathing can hold it square.

Install rim joists early

Normally we sheathe and raise all the first-story walls before installing the second-story floor system. Makes sense, right? But in this case, because we planned to use extra-long sheathing that would extend well above the top plates, we decided to nail the second-story rim joists to the top plate while the wall was still flat on the deck. Joining the rim to the plate "on the flat" is much easier than working off a ladder after the wall is up. Of course, the rims had to be attached to the *outside* edge of the top plate, which introduced a bit of a challenge. Our solution was to support the rim with temporary blocks as we nailed it on.

The rim joists extend to the ends of each wall, creating a slot that the intersecting top plates can slide into later. A triangular square was used to align the ends of the rims with the plates below during installation.

When installing the second top plate, use your hammer to lift it flush with the first plate if necessary.

Make sure the wall is square before sheathing. Once the sheathing is on, the wall cannot be adjusted.

Rim joists are often installed once the walls are up, but it's actually easier to toenail them to the plate beforehand, particularly given the sheathing details required for this house.

Sheathe the Walls

There are a lot of advantages to sheathing a wall while it's still flat on the deck. Gravity will be your friend, and you'll find it easier to align the sheathing to the framing. The technique also prevents the wall from racking as it goes up, which can weaken the joints. But the application of sheathing is one of those techniques that can vary a lot according to the materials being used, the design of the house, local codes, and the preferences of the framer. For example, wood sheathing can be applied vertically or horizontally (the long dimension can go perpendicular or parallel to the studs). In this project, we installed sheathing vertically, but the basic installation technique would be the same if it were running horizontally.

We ran sheathing vertically to create the vertical plywood panels that were required by local code. A *vertical plywood panel* is plywood sheathing that is strategically placed, nailed, and sometimes even glued to the framing according to detailed instructions provided by an engineer. It is typically placed with the long dimension running vertically so that it will tie horizontal and vertical framing members together. These panels increase the ability of the house to resist high winds and earthquakes and are often located at the corners of a house and around window and door openings. If the engineer requires these panels, you'll find detailed drawings on the plans showing exactly where to put them and how to install them (see the drawing below).

Vertical Plywood Panel Locations

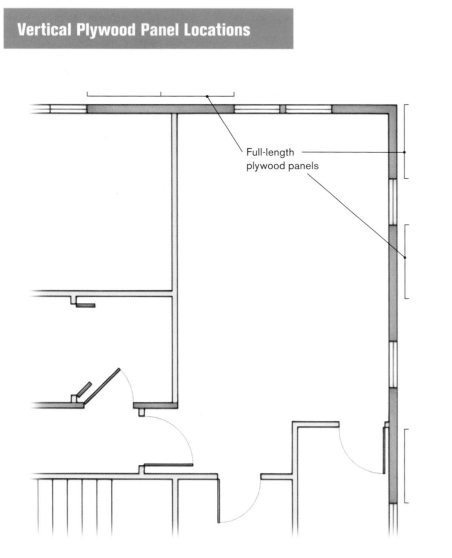

Full-length
plywood panels

ANOTHER WAY TO DO IT

Horizontal Sheathing

When vertical plywood panels aren't necessary, the sheathing can be applied in horizontal runs. With a horizontal pattern, the sheathing is usually left short of the top plate, and the rim joists are installed after the walls are standing. Because horizontal sheathing doesn't cover the entire length of a stud at once, it's easier to tweak the studs and keep them on layout. To install sheathing horizontally, determine how much the sheathing will have to overhang the mudsill, then snap a chalkline across the wall studs 4 ft. from that point. Align the sheathing to the chalkline, with the end seams staggered from course to course so the seams don't line up, and nail the sheathing to the studs. If local codes require it, install blocking to support horizontal seams.

Sheathing must cover the mudsill. To account for the thickness of the floor system, hang the end of your tape over the wall by that amount and mark the length of the sheathing at the other end.

Vertical plywood panels

Panels that reinforce a house against lateral forces such as high wind can take many different forms. On this house; however, the engineer called for continuous vertical plywood sheathing running from the mudsill to at least half way up the second-floor rim joist, the goal being to solidly connect the two floor systems. This requirement meant we had to use 4-ft. by 10-ft. sheets of plywood to sheathe these areas.

Install the sheathing

Any type of wall sheathing should extend past the first-floor rim joists to the bottom edge of the mudsills. Nailing the sheathing to the mudsills provides a structural connection between the walls and the foundation. Measure the amount of overhang from the top of the deck to the bottom of the mudsill (13¾ in. in this case). Then position a tape measure with this distance extending past the bottom plate, and measure up to the 10-ft. mark (see the photo at top). After marking both ends of the wall in this way, snap a chalkline on the rim joist above the wall to serve as a guide for placing the top edge of the sheathing.

After checking the plans to determine where vertical plywood panels were required (see "Vertical Plywood Panel Locations" on the facing page), we marked

Vertical plywood panels must be located precisely according to the plans to maximize the shear strength of the wall. Position plywood sheathing at those places first and fill in the rest of the sheathing later.

those positions on the chalkline. We aligned full-width plywood sheets to the guide line, and tacked them in place at strategic locations. It's faster to nail off the sheathing completely when all the sheets are in place.

Cut openings in place

As you install sheathing, let the sheets extend over door and window openings; it's easier to cut away the

Mark the location of openings on the sheathing before the wall is completely sheathed (above). When the sheathing has been tacked in place, connect the location marks with chalklines (left) or with a drywall square and cut the openings with a circular saw.

Drywall Square

A drywall square is a tool associated primarily with other trades, but framers can use it to improve the speed and accuracy of sheathing installation. It's quite handy for projecting straight lines from the factory edge of the sheathing, which provide guide lines for nailing or cutting.

Raising an Unsheathed Wall

Whenever possible, sheathe a wall before raising it. If you can't, nail temporary 2× braces diagonally across the framing from one corner of the wall to the other. Those braces will keep the wall square as it's being raised. If the wall is small or if it fits between two other plumb walls, bracing may not be required.

plywood later than to cut it to fit beforehand. But be sure to locate the openings before they're completely covered. Tack the sheathing to hold it in place, then measure and snap lines to indicate the location of the openings. Plunge-cut (discussed in the next section) the openings with a circular saw. When the sheathing is in place and the openings have been cut out, go back and nail all the sheathing according to the nailing schedule on the plans.

Before raising the back wall, we framed and sheathed the front wall in much the same way, with one noted exception. A section of the front wall extended past an open stairway and would not have the benefit of floor joists above to stabilize it. That section of the wall also had to support the gable wall of the dormer above, and a rim joist running on top of that section would have been unsupported and weak. To strengthen the wall and to create better support for

the gable wall of the dormer, the crew used longer studs that raised the top plate of that section in line with the second-floor deck sheathing.

Plunge cutting Occasionally you'll need to make cuts with a circular saw that don't begin at the edge of a board or sheet. That technique is called plunge cutting. One example of this procedure is cutting a framed window opening that has been sheathed over.

To make a plunge cut, set the blade at the proper depth for the stock you're cutting and make sure that the area below the cut is clear. Position yourself beside the saw, never behind it. Next, retract the saw guard to

Plunge Cutting

To make a plunge cut, firmly rest the front of the saw baseplate on the work with the blade aligned with the cut line. Turn the saw on and slowly lower it until the saw table is sitting completely on the work.

Continue fastening the sheathing according to the nailing schedule. Snap a chalkline across the wall to indicate the position of the bottom plate.

expose the blade. Rest the front edge of the saw table on the stock and align the blade with the cut line. With the blade slightly above the work, squeeze the trigger. Slowly and carefully lower the blade into the stock while keeping downward and forward pressure on the front of the saw table. When the saw table is flat against the stock, continue the cut to the end of the line. Then reverse the saw direction and cut to the opposite end of the line; *never* pull the saw backward.

Raise, Plumb, and Brace the Walls

Once the front and back walls are completely sheathed, get ready to raise them. A long wall is heavy, especially if it includes extra framing, so make sure it'll stay put once it's up. Nail a 2×4 brace (a stud works fine) at the end of each wall, just over halfway up. Use a single nail so that the brace swings down as the wall goes up, making it ready to be nailed to the rim joist as soon as the wall is upright.

Nail diagonal braces to the ends of the walls. The braces will swing down as the wall is raised, making it easy to brace the wall quickly by nailing through the bottom of the brace and into the rim joist.

Raise the front and back walls

When raising a wall, keep it as flat and even as possible. Even though it is sheathed, a wall can be damaged and weakened if it's allowed to bend excessively. We were lucky on this project to have a fork truck to help with lifting the walls. To make room for the forks, we used a crowbar to lift the top plate of the front wall slightly and slipped 2× blocks under the plates. Then we spread out the crew along the length of the wall and lifted. The fork truck had most of the weight, but the crew helped keep the wall straight as it went up. Even long walls can be raised by hand if you have enough people, but these walls were particularly heavy, so safety was our primary concern. The safest way to raise a wall if you're new to framing is to use wall jacks (see "Raising a Wall with Jacks" below). Most rental outlets that rent construction gear carry wall jacks.

I bet you're wondering what keeps the bottom from kicking out as the wall goes up. Remember those nails you drove into the bottom plate just before layout? They bend as the wall rises, acting as hinges to hold the bottom plate in place. When the wall is standing, the heads of the nails end up under the plates, but the points stick out of the plate and into the floor sheathing. We used a small angle grinder to cut off the exposed nails so they wouldn't interfere with the drywall. A reciprocating saw would work here as well.

On this project, the front and back walls were raised with a little help from a fork truck. Members of the crew make sure that the wall goes up evenly. This is the front wall; the taller portion of the wall is a detail associated with the stairwell.

ANOTHER WAY TO DO IT

Raising a Wall with Jacks

When a wall is long and heavy or if you're working with a just a couple of people, wall jacks are the safest way to go. The bottom of the wall jack must be nailed to the deck, and the lifting bracket is nailed to the top plates. Then you just ratchet the walls safely and slowly into their upright position. Just be sure to lift evenly or you'll twist the wall.

A safe and easy way to raise heavy walls is to use wall jacks. Attached to the deck and to the top plate, wall jacks lift the walls in a slow, controlled manner and prevent them from tipping over.

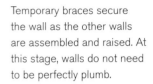

Temporary braces secure the wall as the other walls are assembled and raised. At this stage, walls do not need to be perfectly plumb.

When the wall is up and braced, drive the bottom plate to the chalked layout line. Then nail the plate to the deck at the base of each stud.

To begin layout of the end walls, take measurements from the inside surface of the sheathing on the front wall.

When the wall is standing and approximately plumb, drive a few nails through the end braces and into the first-floor rim joist. Then use a sledge to nudge the base of the wall over to the layout line and drive two or three nails through the plate next to each stud to secure the plate to the deck. Nailing close to a stud is more likely to hit the joist below, but more important, the nails won't be in the way if plumbers and electricians have to drill holes through the plates. Add temporary braces every 10 ft. or so along the length of the wall, anchoring them to temporary blocks nailed through the deck and into a joist below. We raised and braced the back wall the same way.

Raise the end walls

With the first two walls up, the deck is clear for construction of the end walls. The length of the end wall plates is easy to figure—it's the distance between the bottom plates of the front and back walls. Cut plates to length and toenail the bottom plate to the layout line as you did before. Pull your layout by measuring from the outside edge of the framing (the inside surface of the sheathing) on the front wall. That technique ensures that the 4-ft.-wide sheathing will land on a stud when nailed to the end stud of the intersecting wall.

You can build the end walls pretty much as you did the front and back walls, with some exceptions. The L-shaped corner studs aren't necessary because the last stud on an end wall is nailed directly to the corner stud of the wall already in place. Also, you can't put on the entire second top plate until you tip the wall into position. And you can't sheathe the entire wall before you tip it up. For example, the first end wall we built is the one that forms a wall of the garage. That wall extends past the garage just at the ends, so only the ends of the wall would require sheathing. But we couldn't install the sheathing because it needs to lap

With the second top plate complete, toenail any remaining sections of rim joist to the plate.

The wall between the house and the garage went up as bare framing because only the ends would be sheathed. The rest of the second top plate can be installed once the wall is in place.

Add any remaining sections of second top plate to tie the walls together.

over the corners of the walls that are already in place in order to tie the walls together. So we just muscled the bare framing into place. Once the wall was up, we nailed it to the deck and to the corner studs of the intersecting walls.

At the other end of the house, we had to incorporate vertical plywood panels into the framing (wind codes again). So we installed them where we could but didn't sheathe the ends of the wall. After the wall was up, we made sure the bottom plate was on the layout line and then nailed it to the deck.

Complete the top plates

As each end wall is raised and secured, install any missing sections of top plate (see the photo at left). They should overlap the front and back walls and slip into the slots left below the rim joist. Once the walls are tied together, plumb and brace them at the corners.

When plates are in place on all four walls, fill in any missing lengths of rim joists (photo above). Then fill in any missing sections of sheathing.

Garage walls With the main walls complete, you can frame the garage, but it's a little tricky without a nice, flat deck to work on. Our solution was to rest the bottom plate of a wall on the mudsill and support the top plate on blocks so that it sat on the same plane.

We nailed on the sheathing over to the garage to lock the wall in a plumb position at the corners.

After nailing the 2×6 studs in place between the plates, we drilled holes in the bottom plate for anchor bolts.

This is a case in which it really makes sense to hold off on sheathing a wall: The garage walls were light enough to tilt up onto the anchor bolts by hand. We used a prybar to lift each wall slightly until it seated over the bolts. Because the garage walls were raised without sheathing, we took extra care to brace them plumb.

Framing the opening for a garage door is similar to the framing for any door, though there are differences. Garage-door headers can be made from a variety of materials, including dimensional lumber. To make this header particularly strong, we made it out of three LVL layers nailed together. At 1¾ in. per layer, the header ended up 5¼ in. thick, slightly less than the depth of the studs. Because the interior of the garage was not going to be finished, we just made the outside of the header flush with the framing. All garage-door openings are big enough to require multiple jacks, and we doubled the jacks for this one. After framing and erecting the wall, we cut out the bottom plate so it wouldn't interfere with access to the garage.

The garage wall framing was raised before being sheathed, then coaxed onto the foundation anchor bolts.

Garage wall plates and rim joists tie in to the main house rims to complete the garage wall framing.

The second top plate of the garage walls overlaps the top plate of the adjacent house walls, tying them together, and the garage rim joist continues out from the rim joist from the main house. Sheathe the garage walls to complete the first-floor wall framing.

5

Framing the Second-Floor Deck

The house framing version of the classic chicken-or-egg question relates to straightening the walls. Is straightening the final step in wall construction or the first step in building the second-floor deck? I come down on the side of the latter, and here's why. Let's say you finish framing the walls on a Thursday. Friday it rains so you don't work. You're off for the weekend and Monday is a holiday. If you'd straightened the walls on Thursday, they would have had four days to move around in the wind and weather and you'd probably have to tweak them again before framing the floor. The floor framing locks in your straightening efforts, so that's why I associate wall straightening with framing for the second floor.

STEP BY STEP

How to Install the Second-Floor Deck

1 Straighten the walls.
2 Lay out and install the main beams.
3 Lay out and install the house joists.
4 Lay out and install the garage beam and ceiling joists.
5 Complete the stairwell framing.
6 Sheathe the floor.
7 Remove all spring boards.

Prep Work for the Second Floor

Before the wall straightening and deck framing actually started, we built a single interior wall to support the major structural beams for the second floor. I'll discuss the nitty-gritty details of building interior walls in chapter 9. For now, just know that a structural (load-bearing) interior wall must be built just like an outside wall would be and includes structural headers.

Stringing the walls

The walls are standing but they're not necessarily straight. Over the years, I've seen lots of methods, special tools, and jigs for straightening walls, but the springboard method I learned when I first started building houses still works the best in my opinion. Anyone using this method can straighten the walls quickly and without help. It's a two-step process that begins with stringing the walls.

Stringing is the key to perfectly straight walls. Begin the process by nailing 2× blocks to the inside ends of all four exterior walls (see the top photo on p. 104). Then drive two additional nails partway into

each block, placing the top nail at the height of the top plate. These nails act as anchor pegs for the string. Many framers use strong mason's twine for stringing because it can be stretched extremely taut. But a chalkline can work just as well, and it has the advantage of having a hook on the end that can slip over the bottom nail on one block. Lead the string up over the top nail, and then stretch the string as tight as possible to the block at the other end of the wall. At this end, lead the string over the top nail, pull it tight, and then wrap it several times around the bottom nail. Then wrap the string back over itself on the nail to keep it tight. The trick here is not to tie a knot that might have to be untied later.

Structural interior walls are similar to exterior walls, though in this case 2×4s were used instead of 2×6s. Bracing the wall in two planes keeps the wall plumb and square.

Running strings is the first step for straightening the walls. Nail blocks at both ends of the wall, and add nails for pegs (top right). Hook the string onto one end, stretch it, and secure it at the other end (bottom right).

Springboards straighten the walls

Remove the temporary diagonal bracing used to hold the wall upright. (Unless conditions are very windy, even a fairly long wall should stand on its own for now.) The trick is to push the wall in or out to make it perfectly parallel to the string, which is where the springboards come in. For springboard material, we use 12-ft. roughsawn 1×8 boards because they're flexible, strong, and inexpensive. Besides, we can use them for other purposes after they've fulfilled their springing duties.

Walls are usually straightened one at a time, and it really doesn't matter which one you start with. So choose a wall and position springboards every 8 ft. or so along the wall. Nail one end of a springboard to the underside of the top plate. Secure the other end to the deck, giving the board a slight downward bend as you nail it. Then nail a 4-ft.-long 1×8 prop, or kicker, to the

Attach the tops of the 1×8 springboards to the underside of the top plates on each wall (top left). Next, nail them to the deck with a slight bend in them (bottom left). Finally, tack short kickers in place below each springboard (below).

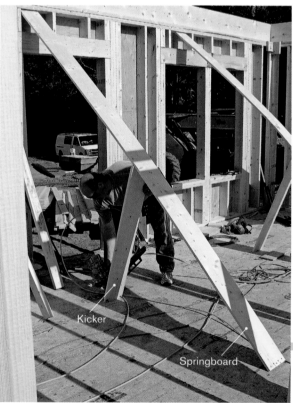

Kicker

Springboard

deck below the springboard. The top of the kicker should be snug against the springboard, but don't nail it yet. Pushing a kicker board in or out changes the amount of arch in the springboard, which in turn moves the wall in or out.

Once the springboards are in place, you can begin to straighten the wall. Slide a gauge block made from a length of 2×4 up to the string with one hand. Push on the kicker until the 2×4 gauge block just slips under the string. Then drive nails through the springboard and into the end of the kicker to hold the wall in place.

After working down the entire length of one wall, sight the plate for a final check, as shown in the photo on p. 106. This is your final opportunity to make sure the wall is dead straight. The slightest deviation in the wall can turn into a major wave once the exterior finish materials are applied.

To straighten the walls, slide a 2× block up to the string as you push on the kicker to move the wall in or out (top right). The string should just graze the edge of the block (middle right). When the wall is in position, drive a nail through the springboard and into the kicker to hold it in place (bottom right).

ANOTHER WAY TO DO IT

A Site-Built Wall Lever

In most cases, springboards can easily straighten a wall. But sometimes more force is required, especially near the end of a wall. A site-built lever lets you apply that force in a controlled fashion. Nail a diagonal 2× brace to a stud near the top of the wall at the trouble spot and nail a long 2× block to the deck next to the loose end of the brace. Now nail a 2× lever to the block and to the brace. Pull back on the lever as someone else gauges the string. When the wall is straight, nail the bottom of the brace to the block to hold the wall in position.

If a wall can't be straightened using a springboard, make a double-action lever to draw the wall over and then anchor the wall with a 2× brace.

As a final check, sight down the plate to make sure it's straight. This is your last chance to get the walls dead straight before installing the second-floor framing.

Setting the LVL Beams

The main framing members of the second-floor deck are dimensional lumber joists. Their arrangement is quite simple: They all run front to back (parallel to the end walls of the house). But unlike the first-floor joists, they don't all sit on top of a support beam—some are framed into the side of a beam. Otherwise, the first-floor walls would have to be higher to create proper headroom below the beam, and higher walls mean more material and more expense. So all the LVL beams supporting the second floor of the house were installed as flush beams. In other words, they're in the same plane as the floor joists. The garage beam is the only one that is not a flush beam.

In general engineering terms, beams 1 and 3 (see "Second-Floor Framing Plan" below) are called

Second-Floor Framing Plan

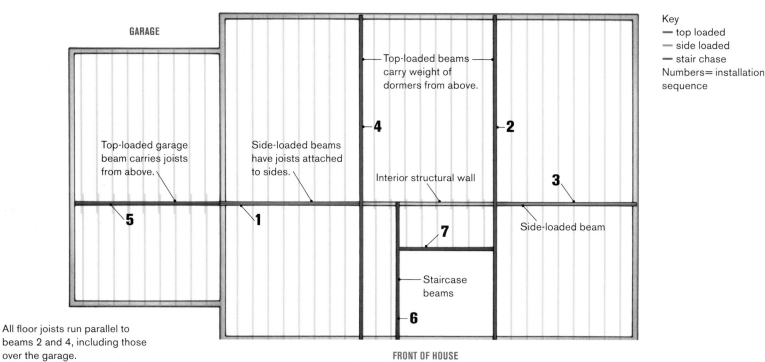

BACK OF HOUSE

GARAGE

Top-loaded beams carry weight of dormers from above.

Top-loaded garage beam carries joists from above.

Side-loaded beams have joists attached to sides.

Interior structural wall

Side-loaded beam

Staircase beams

4

2

5

1

3

7

6

Key
— top loaded
— side loaded
— stair chase
Numbers= installation sequence

All floor joists run parallel to beams 2 and 4, including those over the garage.

FRONT OF HOUSE

side-loaded beams because the load they support (the floor joists) is attached to the sides of the beams. The two beams that run perpendicular to these beams (2 and 4 in the drawing) are called top-loaded beams because the load they carry (the weight of the dormers) sits on top of them. The beam in the garage (5) is also a top-loaded beam because the floor joists above the garage sit on top of the beam. Finally, two small LVL beams (6 and 7) frame the stair opening. For clarity, I've numbered all the beams according to the order in which they were installed.

Lay out the beams

For proper structural support, the ends of the LVLs have to sit directly above solid framing, such as a stud or partition backer, not between studs. In most situations, the beam layout is marked on the top wall plates. But because the rim joists were already in place, we marked the layout directly on the rims and squared the marks down the inside face of the rims, as shown in the top photo at right. Measurements for locating the top-loaded beams come straight from the plans. Having the rims in place actually made it easier to keep the joists and beams in position as we installed them.

By the way, to avoid having to work around the maze of springboards, the crew on this project opted to lay out the beam positions, as well as the joists on the front and back walls, before the walls were straightened.

To locate the ends of beams 1 and 3, measure from the front wall along both end-wall plates. These beams have to be in line with the interior structural wall, so we stretched a string from end wall to end wall to make sure the wall was perfectly in line with the layout.

Install the beams

The LVL beams should be installed in a specific order so that they don't get in the way of the other beams as they go in. These beams can be heavy, so installing them in the proper sequence can save a lot of time and effort—this isn't something you want to do twice. We decided that the side-loaded beam closest to the garage (1) should be the first to go in.

Second-Floor Framing

The framing for the second floor of this house was complicated by the need to provide support for the large dormers, which called for a particular arrangement of LVL beams. In addition, the joists were framed into the sides of the beams to decrease the overall height of the house and maintain adequate headroom on the first floor. Under different circumstances, however, the crew might have framed the second floor with solid lumber joists or I-joists running across the top of one centrally located beam running down the middle of the house.

After determining where the ends of the LVL beams should be, mark the layout on the inside of the rim joists.

A string stretched between the end walls confirms the position of the support wall. The beams must be parallel to the front and back walls and in line with the interior wall.

Short beams such as #1 can be built first and then lifted into place. With no rim joist between the house and the garage on the left end wall, simply toenail the beam to the layout marks on the plate, and toenail the other end to the top of the structural wall.

The long beams #2 and #4 would have been too heavy and unwieldy to handle doubled up, so we set them in place one LVL at a time. Because of their length, we had to feed them through a door or a window opening, then lift them into position on the plates. The danger in assembling a beam in this fashion is accidentally building a curve into it. So before nailing and then bolting the two sides of the beam together, eyeball it and brace it straight.

Beam #2 should go in before beam #4 because it supports another beam (#3). Beam #2 rests on the very end of the structural interior wall, however, leaving no room to support beam #3 with wall framing. To create temporary support during the installation of beam #3, we nailed a block into the end of the structural wall. Then we set the beam on the block and toenailed through it and into beam #2 (see the bottom photo on the facing page). The toenails are simply temporary support; a beam hanger will later be nailed to the longer beam to provide structural support for the shorter beam. Toenail the other end of beam #3 to the rim joist on the end wall.

Beam #1 was assembled before being lifted into place (top) and then toenailed to the plate (above).

It's easy to incorporate unwanted curves when building a beam in place. So before nailing the two sides together, brace the beam to keep it straight.

ESSENTIAL TECHNIQUE

Assembling an LVL Beam

Nails are enough to hold a doubled LVL beam together temporarily, but the manufacturer specifies that LVLs be lagged together with structural lag bolts as well. Whenever possible, the beams should be lagged together before raising, but long beams are too heavy for this and have to be braced straight, then lagged together in place. On this house, lags on the side-loaded beams were added later so their heads wouldn't interfere with the joist installation. Just make sure you get this done before the framing inspection. The crew was able to drive the lag bolts with an impact driver without having to drill pilot holes.

Beams made from multiple LVLs must be lagged together in a particular way. Whenever possible, assemble multiple LVLs before lifting the beam into place.

Beam 3

Temporary block

Beam 2

The long top-loaded beam (2) rests on the very end of the support wall, so a temporary support block was nailed to the end of the wall to hold the intersecting beam (3) temporarily. Later on, a joist hanger will be installed to strengthen the connection.

Long beams are installed in place and then braced straight before being nailed together. The two sides will later be lagged together to form a single, structural beam.

Joists will hang off both sides of this beam, so pull the layout from the wall and extend the layout down both sides of the beam with a triangular square.

To keep the joist layout consistent, extend the layout under the intersecting beams and onto the next beam. Then draw plumb lines at each mark to align the joists on both sides.

The next beam to go in is #4. Pass the LVLs up and over the interior wall one piece at a time, then tap them into position. Brace the beam straight before it's nailed off, then drive lag bolts to hold the beam together.

Joist Installation

With the four major beams in place, mark the joist layout on the side of beams #1 and #3 as well as on the top plate of the structural wall. Where beam #3 intersects beam #2, hold the end of your measuring tape on the last layout mark, extend the tape under the beam, and continue the layout onto the top plate of the

interior structural wall. Then continue the layout onto beam #1. Use a triangular square to carry the layout lines across both faces of both beams as you work.

Area by area

Now it's time for the 2×12 joists. The LVL beams divided the floor neatly into five joist areas: front and back areas on each end of the house, and a center area between them. Although the order of installation really didn't matter, we chose to work right to left, in the direction of the layout, so the joists for the back area on the right side went in first.

To determine the joist lengths, measure the distance between the rim and the beam at both ends of the area and cut all the joists to that length. If the measurements differ slightly but are within 1/8 in. or so, cut all the joists for that area to the shorter length.

Divide the house into areas, and install all the 2×12 joists in one area before moving on to the next.

When all the joists are crowned, cut, and loaded properly, the crown marks should all line up at one end to make installation go more smoothly.

If you find that your measurements differ by more than that, you should really go back and make sure the beam layout is correct. Tweaking the layout now can save a bunch of headaches down the road. Because the joists will be supported by joist hangers nailed to the side-loaded beams, the lengths don't have to be exact. In fact, a little too short lets the joists slip into position easily. Never force a joist into place! It doesn't take much to push a rim joist or beam out of line.

As with the joists for the first-floor deck, crown these joists as you cut them to length (see "Assembling a Beam in Place" on p. 66) and mark the direction of the crown with a large V. For efficiency's sake, crowning, marking, and cutting the joists should be a repetitive process for each section, along with getting the joists into the house and ready for installation. If done properly, all the crown marks end up at one end of the joists facing in the same direction. That makes the installation process go smoothly and efficiently.

Tack all the joists in one area in place (see "A Two-Person Job" at right), and then go back and nail them in completely. Secure one end to the plate and rims, and finish toenailing them to the beam until there are four or five 16d nails on each side of the joist. This might seem excessive because the joists will later be supported by joist hangers. But for now, the toenails

A Two-Person Job

Installing joists always goes more quickly with two people. One person on the ground can set the end of the joist on the plate, then hand the other end to a second person on a ladder. The person on the ladder aligns the joist to the layout mark on the LVL beam, flushes the bottom edges of the joist and the beam, and then toenails the joist to the beam. The person on the ladder drives one nail through the bottom edge of the joist and then a couple more nails through the face on each side of the joist. At this point, those few nails are all that are needed to hold the joist in place.

Use joist hangers to form a secure connection between the joists and the beam. Similarly, install a beam hanger to connect one beam to another.

The joists in the center section rest on the support wall at one end and the exterior wall plate at the other. Roll them up on edge and nail them one by one.

hold everything together until the sheathing can be installed. Make sure heads of the nails are sunk below the surface of the joist so that they don't interfere with the hangers later on. All the hangers must be in place before an inspector will okay the rough framing, but it's often considered a fill-in task left for bad weather or to fill in lag times in the course of the project.

In the center area, one end of each floor joist sits on top of the structural interior wall. These joists can extend beyond the wall as much as you want, as long as they have full bearing on the top wall plate. The joists can be loaded flat on top of the wall plates and

rolled up into position as they're installed. Continue installing each area until all the regular joists for the house are in place.

Garage joists

The second-floor deck continues over the single-car garage. We could have run 2×12 joists across its width, perpendicular to the house joists, with no need for a support beam. But instead, we decided to run the joists parallel to the house joists to keep the layout consistent and so we could continue the floor sheathing uninterrupted from the house into the room over the garage.

Loading Joists

You can load the garage joists flat on top of the wall plate and the LVL beam. The joists lying flat make it easier to walk along the beam and the plate, but be sure to keep one foot over the plate and the other foot close by. Be extremely careful as you step on loose joists to keep them from sliding under foot. The alternative is to set up staging (scaffolding) to work from.

To maintain consistent joist spacing, pull the garage joist layout from the joists in the main house.

That meant we needed a support beam. There was plenty of headroom in the garage, so we installed a top-loaded beam (#5) instead of a flush (side-loaded) beam.

Continue the layout for the garage joists by measuring from the main joists (see the photo above). The length of beam #5 can be determined by measuring from the outside edges of the wall studs on both sides of the garage. The ends of beam #5 sit in fully supported pockets in the garage walls so that the top of the beam is even with the tops of the plates. Be sure there is solid framing in both walls to carry loads from beam #5 to the foundation. If you neglected to put it in earlier, now is the time to do it.

After installing the beam, install the middle garage joists to hold the beam straight. Measure the distance to the beam at each end, and then nail the joists to the same measurement in the middle. The rest of the joists will go in quickly.

Stair-chase framing

The framing for the second-floor deck is now complete except for the area around the stair chase. At this point, we cut out the sheathing that was installed over

the stair hole in the first-floor deck. To locate the edges of the hole, we measured off the walls based on dimensions taken from the plans and snapped guide lines for the saw.

Two intersecting, doubled LVL beams form the perimeter of the stair chase. Beam #6 defines the width of the chase and runs from the front wall plates to the interior support wall. The intersecting beam (#7), called a header, is toenailed to beam #6.

Install the middle joists first to hold the garage beam straight, then install the rest of the joists.

Short joists run from the interior structural wall to the header and then overlap the back joists. Before installing the floor sheathing, we opted to install a first-floor wall to support the stair chase header. We'd have to build it eventually anyway, but by installing it now, the header wouldn't be supported solely by toenailed connections. (Details about building and installing interior walls can be found in chapter 9.)

Another task that's much easier to do before sheathing is installing nailers for the ceiling drywall along the end wall plates. These nailers are 2×6s nailed

flat to the top of the end-wall plates and they're a heck of a lot harder to install after the sheathing is in place. As with the first-floor deck, the plans required blocking between the joists in certain places to pass the high-wind code. When these last framing items are done, the deck is ready for sheathing (and you thought this moment would never arrive).

Cut short joists to fill in the remaining space between the support wall and the stair header.

Until you're able to provide structural support, temporarily support the ends of the stair chase header on blocks nailed securely to the bottom of beams #6 and #7.

Installing an interior wall under the stair header gives it additional support before the floor sheathing is nailed down.

Attach 2×6 nailers to the top plate to serve as backing for the ceiling drywall. This task is much easier to do now, before the floor sheathing goes on.

Second-Floor Sheathing

Sheathing for the second-floor deck happens much as it did on the first-floor deck (see "Floor Sheathing" on p. 73). Snap a line for the first course, and then lay the sheets down starting in one corner. Spread out a few sheets of sheathing on top of the joists to stand on while setting the first course.

As with the first-floor deck, squeeze a bead of glue on the top edge of each joist, and then carefully drop each sheet into place. The LVLs in the second-floor framing did not fall perfectly on layout, so we had to add 2× nailers alongside the beams in a few places to catch the edge of the sheathing.

There are only a few things that differed from the way we installed sheathing on the first floor. For the second floor, we cut most of the sheathing to fit around the stair chase, and we continued sheathing over the garage to form the floor for an unfinished storage room. Once all the sheathing is in place, the first-floor exterior walls are rigid and locked in place. Go back and make sure you've installed all the required joist hangers and beam hangers. Finally, remove the springboards and stack the wood for future reuse.

Sheathing the second-floor deck is mostly a replay of the first floor. Spread construction adhesive on each joist and then carefully drop each sheet into place.

Secure nailers to the sides of the beams to catch the ends of sheathing that do not fall exactly along the center of the beam.

2× nailer

Beware of Bad Sheathing

If you want to install sheathing without a hitch, the tongues and grooves of every sheet have to be nearly perfect. Any broken sections can translate into small scraps that get caught in the groove and prevent the sheets from mating together properly. As you pull sheets off the pile, inspect the edges looking for places where the groove might be damaged. Put those sheets aside to be ripped into strips for finishing off the floor.

Keep an eye out for sheathing with damaged edges. If the groove is crushed like this, you'll never get it to mate properly with the neighboring sheet.

Building Stairs

When I started out as a carpenter, stairs were a huge mystery to me. How do I figure out the angle? How do I know how big to make the steps? Then my good friend and master carpenter, Rob Turnquist, sat me down one day and unraveled the mysteries. Since that day, I've built dozens of stairs, and it's still amazing to me when they fit the way they're supposed to, and it's even more amazing the first time I walk up and down a set of stairs I've built.

In this house, we had an interesting set of stairs to build, but the basic principles we used to design and construct them apply to any straight stairs. Starting from the basement, one flight (a straight section of stairs) rises to a landing, or platform, and then a second flight doubles back as it climbs from the landing to the first floor. The stairs from the first floor to the second floor have the same configuration and stack directly over the basement stairs to use the space efficiently. With this type of stair, the landing is essentially just a big step as far as stair calculations are concerned.

We built the main stairs first to give the crew easier and safer access to the second floor. The basement stairs weren't nearly as necessary so we built them later on. For both stairs, however, we topped the stringers (the diagonal lumber that supports the steps) with temporary treads to stand up to the abuse of dirt, weather, and carpenters' boots. Risers and permanent treads are be added much later when the inside of the house is being finished.

In this chapter, I'll explain how to design and build the main stairs. The basement stairs can be built in pretty much the same way, though I'll note a few minor differences later on.

Calculating Stairs

Humans seem to have a built-in "stair-ometer" that tells us when stairs are comfortable to climb or when they're uncomfortable or even unsafe. It's the same instinct that tells us when we're on a ladder or climbing a hill that's too steep. That sense is a reaction to the ratio of the stairs' overall rise (its height measured vertically) to its overall run (its length measured horizontally). As the slope of the stairs becomes steeper, they start to feel less safe.

Safety also depends on the dimensions of each step being uniform, and inspectors are very particular about this detail. More than a ⅛-in. difference in riser heights can mean that your stairs will fail a final inspection. But more important, that much difference is a safety hazard because it can (and probably will) cause someone to trip as he or she uses the stairs.

Another factor that affects the design of the stairs is the amount of headroom above the treads. The building code requires that the vertical distance

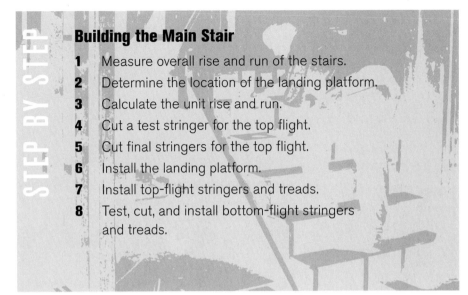

Building the Main Stair

1 Measure overall rise and run of the stairs.
2 Determine the location of the landing platform.
3 Calculate the unit rise and run.
4 Cut a test stringer for the top flight.
5 Cut final stringers for the top flight.
6 Install the landing platform.
7 Install top-flight stringers and treads.
8 Test, cut, and install bottom-flight stringers and treads.

between the tread nosing and any finished ceiling be a minimum of 80 in. This requirement often means making the stairs a little steeper to avoid a forehead-knocking beam, which can be done by shortening the overall run. But make sure the run-to-rise ratio is acceptable and not too steep to be approved.

Yet another factor to consider is whether the stairs are a straight, uninterrupted run or whether they include a landing. In this house, the stairs were designed to fit below the space below a dormer, making the floor plan of the stair chase almost square. The square chase gave the designer more latitude for laying out the rooms on the first and second floors than he would have had with a straight stair run and a narrow, rectangular stair chase.

Measure the overall rise

The first step in designing stairs is to determine their *overall rise*. This is the vertical distance from the surface of the finished floor at the top of the stairs to the surface of the finished floor at the bottom of the stairs. With these stairs, the top of the stairs was directly over the bottom so the measurement was straight up and down. If the stairs are a straight run, you need to level over from the top of the stairs to a point directly over the bottom of the stairs. This strategy gives you the exact overall rise of the stairs regardless of how level the floors are.

For new construction with perfectly level floors, the measurement for overall rise can be taken anywhere. Here, the measurement taken at the back of the chase also locates the height of the landing.

At this stage, the finish flooring isn't in place, of course, but you still have to account for its thickness in your stair calculations. For this house, the finished flooring at both the top and bottom of the stairs would be ¾-in. hardwood, so the overall rise was the same as the distance between the surface of the sheathing on each floor. If 1-in.-thick tile had been specified for the first floor instead of hardwood, for example, that would have shortened the overall rise by ¼ in.

After taking the overall rise measurement, we checked the measurement on the opposite side of the chase and marked the height of the landing platform. Because the floors were dead level, we just measured down from the floor sheathing.

Understanding unit rise and run

The rise of each step, called the unit rise, is the vertical distance between the top of one tread and the top of the next tread. The run of each step, called the unit run, is the horizontal distance from the nosing (the

SAFETY

Working over a Stairwell

To build this set of stairs, the stair chase had to be open. But that left a hole from the basement slab to the second floor. For convenience, we built a temporary platform to support the weight of a single person as the stairs were built. Even if the basement stairs had been built first, they would not have provided an adequate place to stand or to set up a ladder for building both flights of the main stair and the landing. Standard scaffolding would have been problematic to set up. This platform, while crude, was easy to build and covered most of the chase. But if you want to build one that's even stronger, so much the better.

The two-story stairwell had to be left open as stair construction proceeded—there was no room for scaffolding or ladders. To increase their safety, the crew set 2×6s and plywood over the opening to serve as a temporary work platform.

Basic Stair Dimensions

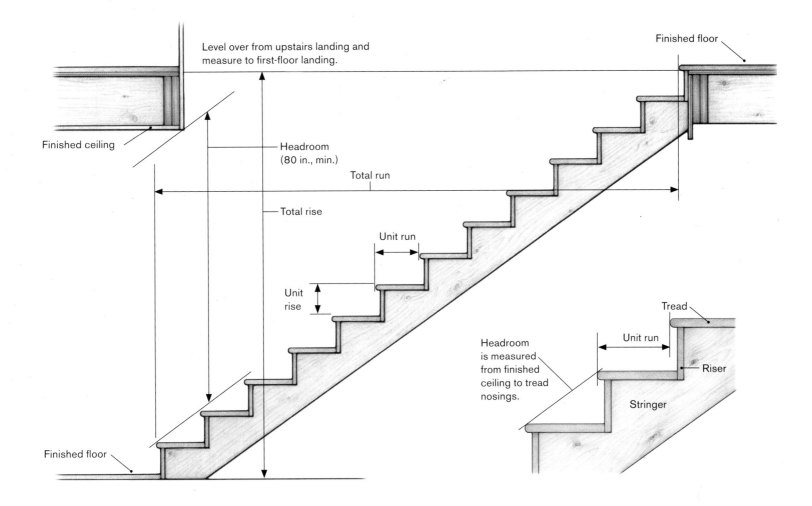

Level over from upstairs landing and measure to first-floor landing.

Finished floor

Finished ceiling

Headroom (80 in., min.)

Total run

Total rise

Unit run

Unit rise

Headroom is measured from finished ceiling to tread nosings.

Finished floor

Tread

Unit run

Riser

Stringer

front edge) of one tread to the nosing of the next tread (see "Basic Stair Dimensions" above). Rise and run are not cast-in-stone numbers, however. To make every step identical and make the stairs begin and end at specific locations, the numbers have to be tweaked. The comfort aspect has to be weighed against other factors, such as tight quarters in the stair chase or limited headroom, to create the safest and most comfortable stairs for the particular situation.

There are many formulas for making sure that the unit run-to-rise ratio falls within acceptable parameters, but each is theoretical. The exact measurements for rise and run vary from staircase to staircase and should never be assumed without careful calculation based on measurements taken at the site. The optimal

step size for stairs is a 7-in. rise and an 11-in. run. But those numbers won't work for every situation, so here are some alternative formulas that are used routinely:

- Rise plus run should equal between 17 in. and 18 in.
- The rise times the run should equal approximately 75 in. ± 3 in.
- Two times the rise plus one run should equal 25 in. ± 1 in.

The last formula is the one I use most often. Note that some communities set the maximum allowable rise at 7 in., so before you build your stairs, check your local codes carefully.

Construction Calculators

Math isn't my strongest suit, so I always use a calculator to figure out stuff like the rise and run of stairs. But if I used a standard calculator, I'd have to enter measurements as decimals and convert the answers back into fractions, a process guaranteed to generate some bizarre numbers. I use a construction calculator instead because I can enter all my measurements in feet and inches, and all the results are displayed the same way. Inexpensive models are available, but for a few bucks more you can get one that can also calculate the length of rafters and other complex house building problems. Even if you use it for only one house, a construction calculator is worth every cent.

- 9 in. + 8⅛ in. = 17⅛ in.
- 9 in. x 8⅛ in. = 73⅛ in. (reasonably close to 75 in.)
- 2 × 8⅛ in. (16¼ in.) + 9 in. = 25¼ in.

Bottom line: The stairs would be slightly steeper than ideal but they would still be comfortable and safe to use.

Cutting the Stair Stringers

Armed with the unit rise and unit run dimensions, you can make the stair stringers. Stringers can be open or closed, as shown in "Types of Stringers" on the facing page. With closed stringers, the treads and risers fit into slots cut into the inside face of the stringer stock. Closed stringers are usually made from finish-grade lumber and are meant to be left exposed.

Open stringers, sometimes called cut stringers, have triangular-shaped notches cut out of their top edge so that the stringer resembles a dragon's tail (a common nickname for open stringers). Open stringers, such as those on this project, are made from framing-grade lumber and are not usually left exposed. The side and underside of these stringers will be covered with drywall. Much later on, the temporary treads will be replaced by finished treads and risers.

Make a test stringer

The first step is to make a test stringer, which will confirm that your calculations are correct. If the stringer is accurate, it will become a template for cutting the rest of the stringers. Choose the straightest, best-quality 2×12 stock to use for stringers. A quick diagonal measurement from the top of the stairs to the platform location marked earlier tells you approximately what length stock to cut the stringer from, but start out with a little extra length just in case. As with floor framing, crown the stringer stock and make sure that any bow faces the direction of the tread and riser cutouts.

Calculate unit rise and unit run

Start your stair calculations by determining the number of rises. This is usually done by dividing the overall rise by 7. But the stairs in this house were not just a simple straight run, so we figured the number of rises a different way. Each stairway actually consisted of two flights: the first from a floor to a landing, and the second from the landing to the floor above. Code specifies that a landing has to be at least 36 in. wide (the minimum width of the main stairs). These stairs were 40 in. wide so we made the platform the same width. Subtracting 40 in. from the width of the chase to account for the landing gave us the overall run and restricted the number of treads for each flight to 6. The overall number of rises would then be 14 because we had to include the rise to the landing and the rise to the floor above.

The overall rise measured 113¾ in. between the first and second floors. To calculate the exact height of a single rise, we divided 113¾ in. by 14. The result is 8⅛ in., which is the height of each riser. A construction calculator really comes in handy for this procedure (see "Construction Calculators" above). Subtracting the width of the landing from the stair-chase opening, we figured the run of each step to be 9 in. While not optimum, that rise and run combination still fit the parameters of our three formulas:

Types of Stringers

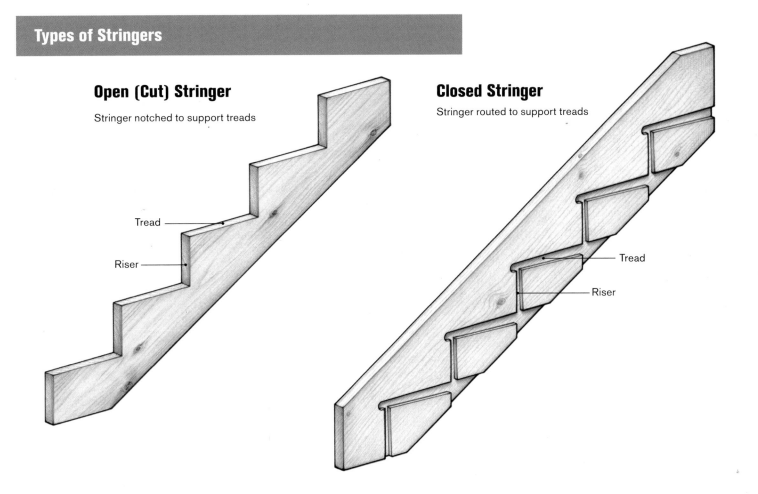

Open (Cut) Stringer

Stringer notched to support treads

Tread

Riser

Closed Stringer

Stringer routed to support treads

Tread

Riser

ANOTHER WAY TO DO IT

Stringer Stock

Stair stringers are typically made from 2×12 lumber. That's because a significant part of the board's width has to be removed for each tread and riser and a 2x12 has enough lumber left after those cutouts. But the problem with dimensional lumber is that it shrinks over time. Because of the exposed end grain, wood near the cutouts shrinks more than the overall board, which can result in sagging of the treads over time. Some builders use LVL stock for stringers because it is less likely to shrink. But pay attention: LVL stock isn't always perfectly straight, so check for a crown as you would check a 2×12.

Stringer Shrinkage

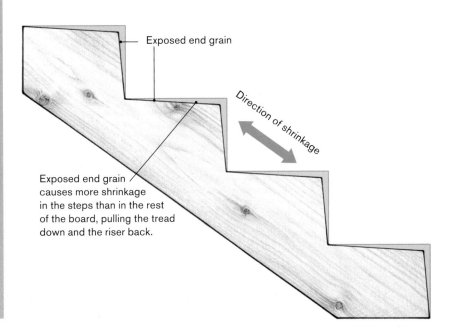

Exposed end grain

Direction of shrinkage

Exposed end grain causes more shrinkage in the steps than in the rest of the board, pulling the tread down and the riser back.

Steel Square and Stair Gauges

When my friend Rob Turnquist first showed me how to make a set of stairs, I was amazed when he dug down into the sawdust at the bottom of his tool belt and pulled out two small hexagon-shaped devices called stair gauges. Each has a slot designed to fit over the edge of a framing square. He slipped the gauges onto the square at the proper rise and run dimensions and tightened the knurled bolts to keep them in place. Stair gauges usually sell for less than $10.

Stair gauges (these two are brass) can be fastened to a steel framing square at the rise and run dimensions. These handy little tools turn the square into a custom stair-layout template.

You'll need a pair of stair gauges for layout. The easiest and most accurate way to lay out a stringer is with a set of these babies and a good old-fashioned steel framing square, not a triangular square (see "Steel Square and Stair Gauges" above).

Set the stringer stock on a pair of sawhorses or a work table with the crown facing toward you. Now tighten a stair gauge at 8⅛ in. on one side of the square to represent the unit rise, and tighten the other gauge at 9 in. on the other leg of the square to represent the unit run. Keep the corner of the framing square flat

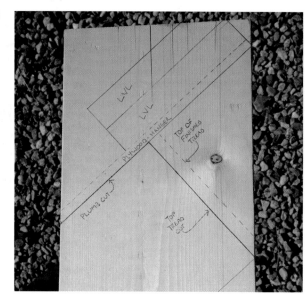

Here is the top of the stringer drawn out along with all the intersecting materials. A full-scale drawing, including the stairwell LVLs, helps ensure that the proper cuts are made from the start.

on the stringer stock, then slide the framing square onto the crowned edge of the board until both gauges touch the edge of the stringer.

Start by laying out the top of the stringer. There are often flaws at the very end of the board that can interfere with the gauges, so place the riser gauge in a bit from the end of the board. Stringers can be confusing, especially to a novice, so make sure you always keep the rise and run on the framing square oriented in the same direction as the progression of your layout. If you stop to check the layout or to count rises or to work from the backside of the stringer, just make sure the square's orientation hasn't changed before resuming the layout.

Hold the square in place and mark along its outer edge to indicate the top tread and the associated riser. Remember that there's no riser at the top of the stringer: The top tread is actually the top of the stringer. Believe me, it's easy to get confused at this juncture. Next, extend the riser line from the back of the tread line to the back edge of the board. This line is the plumb cut where the top of the stringer attaches to the plywood hanger or cleat. By the way, the simplest way to extend riser or tread lines is to place a second framing square or a straightedge against the square with stair gauges.

Slide the square down the stringer stock, and mark out the treads and risers until you've indicated the right number of risers. Number them to minimize confusion.

After establishing the lines for the first tread at the top of the stringer, slide the framing square down, so that the edge of the riser gauge lines up with the tread line. Now trace the next tread and riser. Continue stepping your way down the board until you have the proper number of risers traced out (six, in this case).

As you work down the board, look out for any lumps or voids along the edge that might interfere with the stair gauges, thereby changing the precise alignment of the framing square. (If one of these spots is unavoidable, carefully sight down the milled face of the stair gauge and align it with the edge of the board.) Hold the square down firmly and mark the rise and run on the stringer. At the bottom, extend the tread line below the bottom rise to the back edge of the stock. To make sure the stringers are the right length, I always number the risers and double-check my numbers before cutting them.

The 9-in. run of the treads meant that the bottom riser on the stringers for the upper flight would have only 1½ in. of support on the landing, not nearly

Notches cut in stringers

At the bottom of the stringer, make a full-scale drawing of all intersecting materials (top). The notch at the bottom of the stringer allows it to rest on the landing for support (above).

Basement Stairs Detail

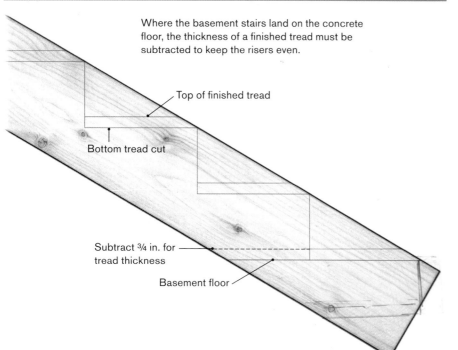

Where the basement stairs land on the concrete floor, the thickness of a finished tread must be subtracted to keep the risers even.

Top of finished tread

Bottom tread cut

Subtract ¾ in. for tread thickness

Basement floor

enough. So at the bottom of those stringers, we made a notch to wrap around the framing of the landing. (On the lower flight, the bottom of the stringers landed fully on the first-floor deck.)

Differences in flooring and tread thickness

One stair-building factor that confounds carpenters and messes up stairs is dealing with the finished floor thickness and the thickness of the finished treads. This house was pretty easy because the finished floors would be ¾-in.-thick hardwood and the treads would also be ¾-in. stock. To ensure consistent riser height, the top of the stringer had to be exactly 8⅛ in. below the second-floor sheathing, or the height of one riser.

But if the treads had been, say, 1¼ in. thick, then we'd have to drop the stringer an extra ½ in., or the difference in thickness between the finished floor (¾ in.) and the tread (1¼ in.). To drop a stringer ½ in., cut that amount off the bottom.

In the basement, the bottom flight landed on the concrete floor with no other flooring. In that case, we took ¾ in. off the bottom of the stringer so that the

distance from the concrete floor to the first step stayed the same as the rest of the steps (see the drawing at left).

So what would you do if the landing were to have a finished floor that was thicker than the treads? Right, you'd drop the height of the landing by the difference in the thicknesses. But in this house, the finished landing floors were the same thickness as the treads, so we set the landing height at exactly seven times the rise, or 56⅞ in. below the second-floor deck. We cut a board that represented one end of the platform and tacked it in place at that height.

Cut just the top and bottom of the test stringer, and set it in place to make sure it fits properly (top). Set a level on one of the tread lines as a final test (above).

Fit the test stringer

Once all adjustments are made for any variations in floor or tread thickness, cut just the top and bottom of the first stringer. This stringer is for testing your layout and landing height. The test stringer for the upper flight had just the plumb cut and tread cut at the top and the notch at the bottom.

Now for the moment of truth. Set the test stringer in place, as shown in the top photo on the facing page. Have someone hold the top at the right height while you check the fit at the top and bottom. In this case, the bottom of the stringer had to be held ¾ in. above the board representing the platform framing to account for floor sheathing on the landing.

You're probably wondering about now why we didn't just build the platform first and then cut stringers to fit. Well, go right ahead and build the platform and then try the test stringer. But when it's not quite right and the platform needs to be taken out with a reciprocating saw to be lowered ¼ in., I'll remind you that it's a lot easier to tweak the height of the platform when it's represented by a single board nailed to the framing. That's what the pros do to save themselves time and trouble in the long run.

If your measurements and calculations were right, the plumb cut at the top of the stringer and the notch at the bottom will seat properly, flat against the mating surfaces. Set a torpedo level along one of the tread lines to confirm that the steps made from this stringer will be level.

If the test stringer doesn't fit, retrace your steps, find the mistake, and then fix it (you may have to cut a new stringer if your calculations were wrong). Take time now to make sure the stringer fits exactly. Correcting the problem later is guaranteed to be a pain.

When you know the stringer will fit, cut along the rise and run layout lines. Then go back and finish the cuts with a handsaw.

Use the test stringer as a pattern

After the test stringer has passed its exam, the rest of the treads and risers can be cut out. Many carpenters just cut beyond the intersecting rise and run lines so that the triangular piece of stringer falls away, but overcutting the stringer weakens it. Instead, cut just to the lines and finish the job with a handsaw. The amount of extra work this takes is minimal compared to the benefit it brings.

When all the cutouts are made on the first stringer, you can use it as a pattern for the other stringers in the same flight. Simply align the bottom edge of the pattern on the stringer stock and trace every detail (see the top photo on p. 126).

Use the first cut stringer as a pattern for laying out the rest of the stringers in a flight.

How Many Stringers?

The number of stringers needed for a staircase depends on the treads: They have to span the distance between stringers and you don't want them to flex. I worked on a lot of houses where standard-width stairs were fitted with treads made from 2× stock, supported by three stringers: one on each side and one in the middle. (Though I noticed that longer stair runs got a little springy sometimes.) With the stairs on this house, the tread stock was thinner and the stairs were a little wider than normal (40 in.). So we built the stairs using four stringers. The extra one made these stairs really solid.

Installing the Stairs

A piece of ½-in. plywood nailed to the stair chase framing can be used to attach the stringers. Note that the floor sheathing has not been trimmed back in this photo.

There are several ways we could have attached the top of each flight to the platform or the floor framing. For example, stringers can be notched to hang on a 2× cleat or can fit into a metal stair bracket. We chose to use a ½-in.-thick plywood hanger instead. The hanger is as wide as the stairs and tall enough to capture the plumb cuts at the tops of the stringers. This method has a couple of advantages. Nailing through the front side of the plywood secures the hanger to the house framing. After that it's easy to nail through the *back* side of the plywood and into the ends of the stringers. Using a plywood hanger also enables you to install one stringer at a time. Measure for the hanger when the test stringer is in place to be sure it will be wide enough to catch the ends of the stringers.

Once all the stringers for the top flight are complete, set them aside and build the landing platform. Because the landing is much smaller than the main floors, it can be framed with 2×10s. The perimeter framing was nailed directly to the studs, and the 2×10 that carried the ends of the stairs extended beyond the platform into the wall framing where it was supported by a 2× post.

Thinking Ahead

Here's something that will make your life easier well after the framing is done, when it's time for interior details. Nail a 2×4 along the bottom edge of the outside stringers. The 2×4 spaces the outside stringers away from the wall so that wall finish and skirtboard trim can slip past the stringers without having to be cut to fit the dragon's tail precisely.

The 2×4 spacers create a gap beside the stringer, which eliminates a lot of finicky cutting and fitting when it comes time to install drywall and finish trim.

2× spacer

Attach the landing platform to three walls of the stair chase. Support the open side of the platform by extending the framing over 2× posts in the walls at each end.

Install the top flight

The next part is easy: nailing the stringers into place. Space the stringers evenly across the width of the stairway. These stairs were 40 in. wide, and the stringer stock along with the spacers totaled 9 in. Four stringers created three spaces, so divide the remaining 31 in. by three. The spacing just has to be close, not exact, so we made the two outside spaces 10¼ in. wide and the middle space 10½ in. wide.

Toenail the top of the stringer through the plywood and into the LVL beam that frames the opening. Where the stringer hangs down below the LVL, drive nails through the back of the plywood hanger and into the ends of the stringers (as shown in the top photo on p. 128). Nail the bottom of each stringer through the notch and into the platform.

To attach the stringers, nail through the plywood hanger and into the plumb cut of the stringer. The top of the stringers can be toenailed into the floor framing.

Line up the temporary treads with the front and outside edges of the stringers, and nail them in.

As with the top flight, set one bottom stringer in place to make sure it fits properly. Then use that stringer as a pattern for the rest of the bottom stringers. The bottom edges of these stringers land on top of the first-floor deck sheathing.

Install the temporary treads

Remember all those springboards we used to straighten the walls back in chapter 5? I told you they'd be put to good use: Cut them up to serve as temporary stair treads. A lot of carpenters put the temporary treads in haphazardly. But if you're careful and make them flush with the outside and front edges of the stringer, the stairs will be easier to use for the rest of the construction process, and there won't be any overhanging treads to catch hoses and extension cords.

Install the bottom flight

The bottom flight of stairs uses the same rise and run dimensions as the top flight. The overall layout is the same, too, except that the bottom of the stringer lands fully on the first-floor deck, which is actually preferable to being notched. Make a test stringer as before and check its fit. Use that stringer as a pattern, and then cut the rest of the bottom-flight stringers.

Nail the top of each stringer to a plywood hanger and toenail the other end directly into the first-floor deck. The final test (and my favorite part) comes after the last tread is in place. It's a great feeling to walk up and down a set of stairs for the first time.

Basement stairs

We held off on installing the basement stairs for a while because we just didn't need them right away. Design and construction followed pretty much the same sequence as the main stairs, but with a few differences. A crucial part of laying out these stairs is remembering to subtract the thickness of the tread from the bottom of the stringers (see "Basement Stairs Detail" on p. 124). Where the stairs rest on the concrete floor, it's also a good idea to cut an extra 1½ in. off the bottom of the stringers. Then anchor the stairs to the concrete with a pad made from pressure-treated wood, so that the stringers won't absorb moisture from the concrete and rot. I make the pad out of a 2×6 or 2×8 so that the whole bottom of the stringer bears on it. Attach the pad to the floor with masonry nails or powder-actuated fasteners.

The ultimate test for stairs is the human test: walk up and down to make sure the steps are safe and even.

SAFETY

Stair safety on the job site

Stairs are a wonderful convenience on the job site. It's so much easier to lug tools and materials up a stairway instead of a ladder. But to make them safe during construction, build temporary railings around the stair opening and at the open side of the stairs to prevent accidental falls.

Because we didn't build the basement stairs right away, we blocked off the open hole with plywood. When the second-floor framing got under way, railings made from 2×4s and plywood were built around the second-floor stair opening.

During construction, block off the open stair chase for safety. The stairs to the basement were built later, so the chase was covered with plywood.

Framing Gable Walls

I have always thought it interesting that a home is referred to in general terms as "a roof over your head." Not four walls around you or a foundation under you. No, the thing that defines a home is that roof. It keeps out all manner of weather and completes the enclosure that creates a house.

Like stair building, roof framing requires calculations. But once you get the numbers straight, building a roof can actually go pretty quickly. I've worked with crews who frame and sheathe a main roof before starting the dormers (this house has two), and with crews who frame them simultaneously. In this case, the crew built the gable walls for the house *and* the dormers first, then installed all the ridges, rafters, and sheathing.

Gable Wall Anatomy

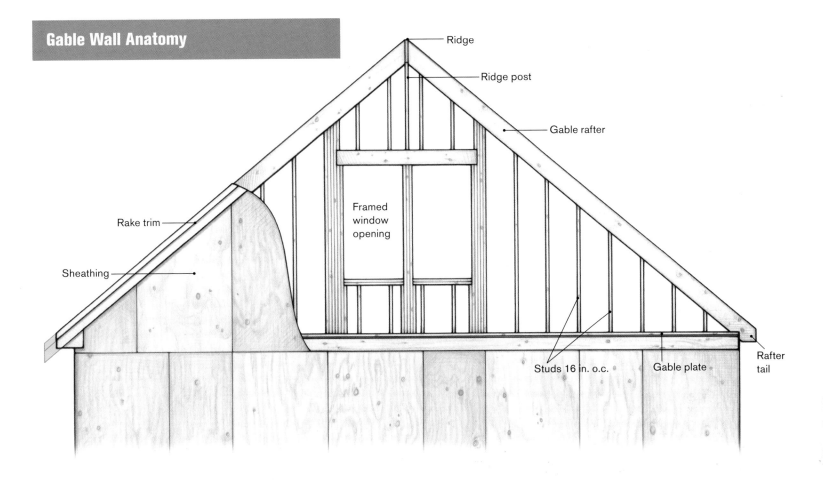

Ridge

Ridge post

Gable rafter

Rake trim

Framed
window
opening

Sheathing

Studs 16 in. o.c.

Gable plate

Rafter
tail

This is a lot of material to cover so I'm going to tackle it in two chunks. In this chapter, I'll show you how to build gable walls and how to lay out a rafter. Once the gable walls are in place, we installed rafters and ridges between them to complete the roof. So in chapter 8 I'll cover ridges, more rafter work, and roof sheathing. Once the roof is framed and sheathed, the house will finally offer its first modicum of shelter.

Installing the Main Gable Wall Plates

There are lots of ways to frame the top story of a house and its roof. For this project, the rafters rest atop plates attached to the second-floor deck to reduce the overall height of the house. Other strategies that involve placing the rafters on top of second-story kneewalls or full-height walls raise the overall height of the house significantly.

The gable walls are those triangle-shaped walls at the ends of the house. Each one is actually a combination of wall studs and rafters (see "Gable Wall Anatomy" above).

Square the deck

Framing almost always begins when somebody reaches for a chalkline, and it's no different here. Snap lines on the deck to guide placement of the gable wall plates, then snap lines for the rafter plates (plates that

STEP BY STEP

Gable Wall Framing

1 Cut and set the main gable wall plates.
2 Lay out and cut the main gable pattern rafter.
3 Build and raise the main gables.
4 Build and raise the dormer gables in similar fashion.
5 Build and install the cheek walls.

Toenail the gable wall plates to the deck along the chalked layout line.

Metal straps nailed to the bottom of the plate and to the deck act as hinges and restraints to keep the base of the gable in place as you raise it.

the lower ends of the rafters will sit on). Depending on how fast you work, it's not a bad idea to snap the lines in permanent chalk. This is one of the last chances Mother Nature has to wash away chalklines. To snap the plate lines, measure in 5½ in. from the edges of the deck near each corner of the house (once again, remember to plumb up from the rim joist instead of just hooking your tape on the edge of the sheathing). Then mark those points and snap lines between them. Take diagonal measurements between opposite corners to make sure the layout is square and adjust the lines if necessary. At the risk of sounding like a broken record, let me reemphasize the importance of making the plate lines absolutely square to each other. Why? Because, as the saying goes, "stuff happens." It could be a rim joist that's thicker than the one it butts to or maybe a bunch of slight imperfections that combine to throw the deck off square. At each new phase of construction it's important to reestablish square because any deviation makes it harder to build the roof.

Cut and set the gable plates

Some carpenters build gable walls with top and bottom plates (see "Gables with Top Plates" on p. 134). However, the gable walls in this house had only a bottom plate because the tops of the wall studs were nailed directly to the gable rafters. Lines snapped at the ends of the house are the guidelines for these bottom plates. Toenail a straight 2×6 to the snapped line, as shown in the left photo above. When the first length of plate is secure, install the next one and continue across the length of the wall.

Gable walls are almost always taller, heavier, and more unwieldy than regular house walls, so take precautions to prevent the bottom of the wall from sliding off the deck as you lift it into place. Here's a cheap and low-tech safety solution that works beautifully. Cut an 18-in.-long piece of metal binding strap from a bundle of lumber and nail it to the underside of the plate and to the deck. Put a strap every 8 ft. or so along the plate, but remember to cut off the ends of the straps once all the framing is complete.

Rafter Length and Run

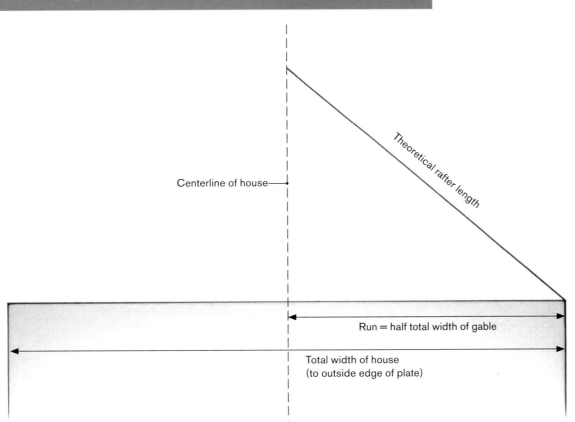

Centerline of house

Theoretical rafter length

Run = half total width of gable

Total width of house
(to outside edge of plate)

Measure and Cut the Gable Rafters

Determining the proper length of a rafter has left many a carpenter scratching his head. The plans give you the roof pitch and the total run of the rafter, but you have to figure out the actual length of the rafter from that information. Pitch is a measure of the angle (slope) of a roof. It is a ratio of the height that the roof rises per 12 in. of run. The roof on this house was a 10 pitch, which is written as 10:12 or 10-in-12.

Determine the length of common rafters

Common rafters are rafters that are at right angles to the ridge of a house and go all the way to the eaves.

They're the simplest rafters to calculate and cut, and they're used for the main roof of this house and most of the dormer roof framing.

Different framers calculate the length of a common rafter differently, but in any case it's a matter of imagining some portion of the rafter as the hypotenuse of a triangle, then calculating the rafter length from that. I think of the hypotenuse as a line that runs from the centerline of the house at the ridge, to the outer edge of the plate (see "Rafter Length and Run" above). That measurement is called the *theoretical* rafter length. Once I have that figure, I add the length of the rafter tail (if any). The tail forms the eaves (the part of the roof that hangs over the side of the house). Then I subtract a small amount that accounts for the thickness of the ridge. The resulting figure is the actual length of the rafter that I'll cut. Trust me, this will all become clear later.

Gables with Top Plates

On a gable wall with top plates, those plates follow the angle of the roof just below the rafters. The top of each gable wall stud must be beveled to fit against the sloping plate. The advantage is that the plate can be nailed to the studs across their entire width, which in theory gives the studs better bearing and offers more support for the rafters. The drawbacks are the cost of extra wood for the top plates and the extra time required for layout. Rafter layout is the same whether or not the gables have top plates.

An alternative method for framing a gable wall is to cap the wall with top plates. The rafters can then be toenailed to the sloping plates.

Top plates

First you need to get the measurement from the centerline of the ridge to the edge of the plate, which is the run of the gable-end rafters. This house has a symmetrical roof, so the run of a single rafter is one half the width of the house. When the gables were framed, the top layer of sheathing had not been put on the walls, so we measured the width of the house from the outside of the framing on both sides. Just because the plans say the house should be 28 ft. wide, however, don't assume that the run is 14 ft. In our case, the actual width of the house was 28 ft. ¼ in., so the run was 14 ft. ⅛ in.

You bought a good construction calculator when you built the stairs, right? Good, because now it'll *really* pay for itself. Simply enter the roof pitch and the run of the rafter, and the calculator does the rest. For this house, the theoretical rafter length (the length of the rafter before doing any additional calculations) was 18 ft. 2⅞ in.

Lay out and cut the pattern rafter

Like a test stringer for the stairs, the pattern rafter confirms your calculations and serves as your template for making the rest of the rafters. A typical rafter has a plumb cut at each end and a bird's-mouth cut where the rafter sits on the plate (see "Detail of Rafter Tail" on p. 136). Once you know the theoretical length of the rafter, you can figure out how to lay out the bird's mouth and how to detail the rafter tail.

Crowning the rafters Now you can shift into rafter-preparation mode. Go through the rafter stock, crown all the boards, and stack them with the crown going in the same direction. Crowning is a crucial step because the roof plane is highly visible from the ground. I once built a garage on a house I owned. Friends helped me cut the rafters, and one rafter did not get crowned properly. I didn't catch the mistake until the roofing was on: There was a distinctive dip in the roof plane over that rafter, and it bugged me as long as I lived there.

Calculating Rafter Length

To find the length of rafters, you can use a standard calculator and rafter tables instead of a construction calculator. Rafter tables are charts that can be found in books, online, or on the side of a steel framing square. Some tables are written as length per inch of run; others (such as those on the framing square) are written as length per foot of run. If the length of the run includes a fraction of an inch, converting your measurements to length per inch works best. Or you can divide the length per foot by 12.

Steel framing squares are covered with numbers, and I admit that I don't have the foggiest notion of what some of them represent. At least the parts referring to rafters are labeled. The line we're concerned with is "Length Common Rafters Per Foot Run." Follow this line down to the 10-in. mark (that represents our 10:12 pitch) and just below it you'll see 15.62. Divide 15.62 by 12 to get 1.3017 in. per inch of run in a 10:12 pitch. Now multiply the length per inch by 168.125 in. (14 ft. 1/8 in.) to get the theoretical rafter length of 18 ft. 2 7/8 in.

There are other ways to determine rafter length, too, and many professional carpenters lay out rafters every day with nothing more than a framing square and stair gauges. But I prefer using rafter tables or a construction calculator.

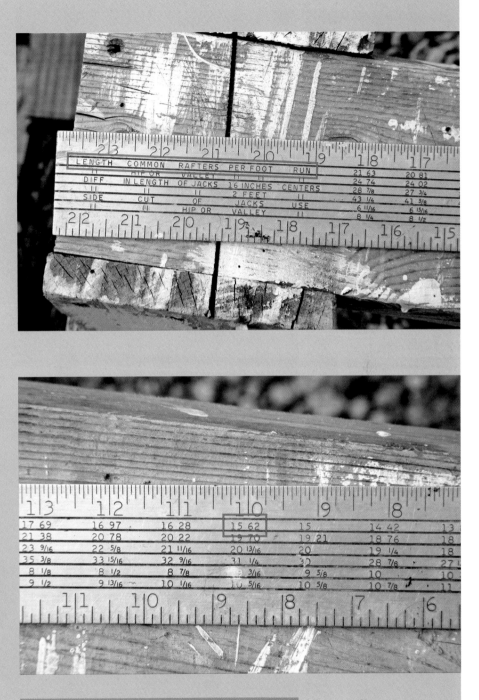

Every framing square includes a list of tables on one side of the square. To find the length of a common rafter with a 10:12 pitch, follow the top line to the 10-in. mark. With that length (15.62 in.) and the width of the house, you can determine the length of the entire rafter.

Detail of Rafter Tail

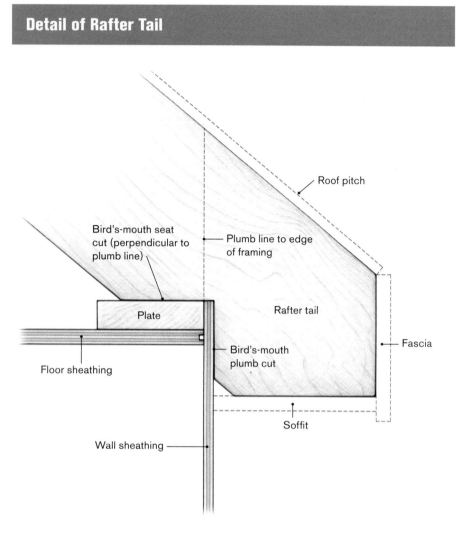

Roof pitch

Bird's-mouth seat cut (perpendicular to plumb line)

Plumb line to edge of framing

Plate

Rafter tail

Floor sheathing

Bird's-mouth plumb cut

Fascia

Wall sheathing

Soffit

mark the theoretical length. Draw another plumb line at that point to indicate the outer edge of the gable plate (see "Detail of Rafter Tail" at left). The width of the rafter tail is measured horizontally from that line as well. When these gables were being built, a band of wall sheathing still had to be patched in along the tops of the walls, so next we measured over ½ in. and drew another plumb line to indicate the outside of the sheathing. If you don't account for the sheathing now, you'll have to do a lot of cutting and piecing to fill in the wall sheathing around the rafter tails. (If the sheathing had been in place, the run of the rafters would have been calculated from the outside of the sheathing, rather than from the outside of the plate.)

Cutting the bird's mouth The line for the seat cut comes next. It is the horizontal line of the triangular cutout called the bird's mouth. The vertical line for the bird's mouth is the outermost plumb line, in this case, the line we drew for the wall sheathing. Flip the square over and place it against the bottom edge of the board. The seat cut is perpendicular to the plumb cut, so the angle of the seat cut is complementary to the plumb cut angle. We earlier established that the 10:12 pitch angle is 40 degrees, so the complementary angle is 50 degrees.

Making the top plumb cut Choose a straight 2×10 for a pattern rafter. Start by marking and making a plumb cut at the top of the rafter. A large triangular square works great for this step. Hook the lip of the square on the top of the board and swing it until the 10-pitch mark lines up with the top edge of the board, then mark the plumb cut down the adjacent edge of the square and cut along the line. (Note at this point that the corresponding angle for a 10:12 pitch is 40 degrees. You'll need that info in a moment.) By the way, stair gauges set at 10 in. and 12 in. on a steel framing square work just as well. Go ahead and make the plumb cut.

Marking the bottom plumb cut Now hook your tape on the peak of the plumb cut and measure along the top edge of the rafter toward the other end, and

To account for the sheathing on the walls below, move the square ½ in. (the thickness of the sheathing), and draw a second plumb line.

To draw a line for a plumb cut at the top of the rafter, pivot the triangular square to a 10:12 pitch (second set of numbers in from the edge) and trace along the edge.

Swing the square to 50 degrees and you have the seat cut angle (see "Detail of Rafter Tail" on the facing page). We set the length of the seat cut at 4 in.

Calculating the height above plate The bird's mouth removes a triangular chunk of the rafter. As the pitch changes, so does the amount of material taken out of the board if you keep the seat cut at the same length. To make sure the bird's mouth does not weaken the rafter excessively, you have to make one more calculation to determine the height above plate (HAP). This is a measure of how much rafter material is left after cutting out the bird's mouth.

The HAP is measured from the corner of the bird's mouth either squarely across the rafter or along a plumb line that extends upward from the bird's mouth (see "Height above Plate" at right). The rule of thumb is that the HAP should be no less than two thirds of the rafter width. A 10:12-pitch plumb line across a 2×10 rafter is 12 in. long, so the HAP should be at least 8 in. (two thirds of 12 is 8). Squaring *across* the same rafter, the HAP should be at least 6⅜ in. (two-thirds of 9½ in. is 6⅜ in.). If the HAP is less than two-thirds, lower the seat cut accordingly. (This raises the height of the ridge by the same amount.)

Height above Plate

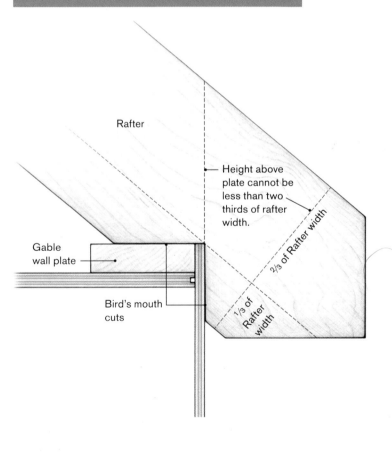

Rafter

Height above plate cannot be less than two thirds of rafter width.

Gable wall plate

Bird's mouth cuts

⅔ of Rafter width

⅓ of Rafter width

By laying out the bottom end of the rafter and labeling all the parts, you can double-check all the calculations made earlier.

At the top of the rafter, subtract half the thickness of the ridge, or ¾ in. for a 2× ridge.

Laying out the rafter tail Now you can lay out the rafter tail. I draw the whole rafter tail full scale on the stock to see exactly how all the eaves details are going to work. For this house, the soffit, or the horizontal underside of the eaves, consisted of a continuous vent with strips of wood on either side. The width of that detail plus a little wiggle room came to 8½ in. So next we measured that distance from the plumb line and drew another plumb line. The final line for the rafter tail is the soffit line. We set the height of the soffit based on the size of the fascia trim and the reveal, or the amount that the fascia extends past the soffit.

Making the ridge adjustment The final adjustment to the length of the rafter accounts for the thickness of the ridge board. The ridge on this house is a 2×12, so subtract ¾ in. (half its thickness) from the plumb cut at the peak of the rafter. Simply draw a second plumb line parallel to the first one and ¾ in. away. This is sometimes called the ridge allowance.

At the top of the rafter, use a circular saw to cut along the second plumb line. Cutting the bird's mouth and the overhang detail at the rafter tail completes the pattern rafter. This will serve as your guide to laying out the rest of the common rafters in that roof section.

This house actually calls for making three different common rafters, and we had to make a pattern rafter

for each: for the main house, the garage, and the dormers. To keep everything straight, make sure you label each pattern with the run of the rafter, the theoretical length, and the pitch.

Build and Raise the Gables

Use the pattern rafter to trace two rafters for each gable and then cut them. Set a pair at each end of the house with the gable plates. Before nailing anything and before cutting more rafters, make sure these two rafters fit properly at the plates and at the peak of the gable. If they don't, figure out why and make corrections. You may have to recut your pattern. If you're working with a crew, you can assemble both gables at the same time; otherwise build and raise them one by one.

Temporary 2×6 spacers nailed to the edge of the rafters hold them at the proper height as you assemble the gables on the deck.

Nail the bottom plate of the gable to the seat cut of the bird's mouth at the bottom of the rafter.

Nail in a temporary block as a placeholder for the ridge. A second piece of 2×6 holds the rafters together until the wall is framed and the sheathing is nailed on.

Build the gable triangle

Like the rim joists we installed on top of the first-floor walls, the rafters should be nailed to the upper face of the gable walls so that they'll be in the same plane as the wall when the gable is raised. To hold the rafters up at the right height as you assemble the gable, nail a couple of 2×6 spacers to the edge of the rafters about 4 ft. from either end. Then nail the bottom of each rafter to the gable wall plate with two 16d nails; be sure to allow space for the sheathing. Where the two rafters meet, tack an 11½-in.-long block of 2×6 flush to the top of the rafter as a place holder for the 2×12 ridge. An additional scrap of 2×6 bridges the ridge block and the rafter peaks to hold the assembly together.

Add the wall framing

Both of the main gables on this house had an opening centered on the wall. The right gable had a double window, and the left gable had a door for access to the room over the garage. To help lay out these openings, snap a centerline on the deck from the midpoint of the bottom plate to the center of the ridge block (see the left photo on p. 140).

Openings in gable walls are laid out and assembled before the rest of the wall, just like regular walls (see chapter 4). However, every stud in a gable wall has to be cut to fit under the rafters, so each stud is a different

A line snapped from the center of the plate to the middle of the ridge block helps guide the layout of the door opening in the gable.

length. Start with the king studs. Locate their position on both the bottom plate and on each rafter by measuring to each side of the centerline half the width of the rough opening plus the thickness of the jacks. Next, measure from the king-stud layout point on the plate to its layout point on the rafter. That measurement equates to the long point of the pitch cut that will let the stud fit against the bottom edge of the rafter.

Gable studs get a notch (sometimes called a flag cut) at their top end, so that part of each stud slides under the rafter; an angled cut seats against the bottom edge. To lay out a flag cut, mark the length of the stud and then draw the cutline at a 10:12 pitch using a triangle square. Set the depth of your circular-saw blade to 1½ in. and cut along the angled line (see the drawing on the facing page). Then lay the 2×6 down flat and make a rip cut to meet the first cut. It's a good idea to clamp the board securely when making these cuts, particularly the rip cut, to keep your hands well away from the action. The flag part doesn't have to be an exact length, but cut it so that the flag doesn't

Measuring over from the snapped line to the rafter sets the width of the rough opening, and measuring from the plate to that point is the length of the king stud.

extend beyond the top of the rafter. If you do it right, the flag will slip neatly under the rafter.

Nail in the king studs, jacks, and headers for the openings, then mark the stud layout on the plates, measuring from the front wall. Measure from the king stud to the nearest stud layout at the plate, and then mark that same distance from the king stud to the rafter. Measure between those points on the plate and on the rafter to get the stud length. When the first stud has been cut and installed, hook your tape on it and mark the rest of the stud layouts where the measurements intersect the bottom edge of the rafters. With the layouts marked, go back, measure the length for each stud, and make a cutting list. Then you can cut all of the flag-top studs at once.

One word of caution: Make sure you indicate whether the measurement is to the long point or to the

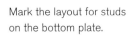
Mark the layout for studs on the bottom plate.

The flag end of the stud slides under the rafter, and the pitch cut seats along the edge of the rafter.

Cutting a Flagged Stud

After measuring and installing the first stud, pull measurements to the rafter for the other studs. Then measure the length of each one and cut them all at once.

short point of the diagonal cut. As you pull your layout across the middle of the gable, the marks shift from being the long point to being the short point, which can be confusing. I'd like to have a dollar for every time I've made *that* mistake.

One of the most important components of gable framing is the ridge post: a stud or partial stud that supports the ridge. The ridges in this roof are considered nonstructural (see chapter 1), but the weight of

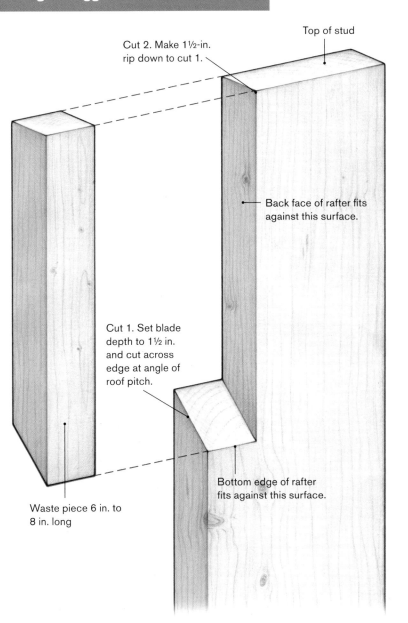

Top of stud

Cut 2. Make 1½-in. rip down to cut 1.

Back face of rafter fits against this surface.

Cut 1. Set blade depth to 1½ in. and cut across edge at angle of roof pitch.

Bottom edge of rafter fits against this surface.

Waste piece 6 in. to 8 in. long

the ridge framing itself needs solid support. Structural headers over the openings support the ridge post.

Install the wall sheathing

The sheathing for the two main gables was not the same. The right-hand gable was completely an exterior wall, but the left-hand gable was overlapped by the garage roof. In that case, only the portion of the gable exposed to the exterior required sheathing.

Right-hand gable Before sheathing the gable, remove the temporary block for the ridge. On the right-hand gable, the courses of plywood had to be installed vertically on the wall to continue the pattern from the first-floor walls. Because first-floor wall sheathing ended halfway up the rim joist, the gable sheathing had to extend past the bottom plate to meet it. It's not important that first-floor sheathing meet the second-floor sheathing exactly, so make the sheathing ¼ in. short to ensure that it doesn't hit the lower

sheathing when you tip the wall into position. Nail the sheathing using the same nails and same nailing schedule you used on the first floor. When nailing sheathing to a gable wall, let it run past the edge of the rafter and tack it in place. Then snap a line to indicate the top edge of the rafter and trim off the excess sheathing.

Left-hand gable The garage roof ends at the left-hand gable. So before sheathing the gable, lay out the gable and rafters for the garage roof. As you did for the main roof, make a pattern rafter for the garage and use it to make two pairs of garage rafters.

After test-fitting them, set one pair of rafters on top of the framing for the left gable of the main house. Measure the distance that the garage steps in from the main house and mark that distance on the plate of the left gable. Use that mark to align the bottoms of the garage rafters. Trace the bottom edge of the garage rafters on the gable studs. This line indicates the lower

The ridge will be supported directly by the ridge post. In this case, a header supports the ridge post. If there were no opening in the wall, the ridge post would extend all the way to the gable wall plate.

The wall for the right gable was sheathed vertically like the walls below. Let the sheathing hang over the rafter, then snap a line and cut off the excess after the sheathing is nailed.

Only part of the left-hand gable has to be sheathed because it coincides with the garage gable framing. Position garage rafters against the house gable and trace along the bottom edge. Sheathe down to the line, then nail the garage rafters on top of the sheathing.

Garage rafters

The strips act as spacers so that the siding material can slide behind the rake trim.

The rake trim detail for this house was about as simple as it gets: a 1×8 topped by a 1×4. Start by making a 10:12 pitch plumb cut at the top of a 1×8. We used stock that comes preprimed so that the boards are protected on all sides and edges. I've even seen some builders put a coat of finish paint on the trim before the gable is raised. Align the plumb cut perfectly in the peak of the gable, but instead of keeping the top edge flush with the framing and sheathing below, space the board ½ in. beyond the framing (see the top photo on p. 144). The extra space keeps the edge of the roof sheathing hidden behind the trim when it goes on. The heads of the nails that fasten the trim boards will be exposed to the weather, so attach all the trim boards with stainless-steel nails.

A 1×3 furring strip spaces the rake board away from the gable. This strip creates a space for the siding to slip behind the trim.

edge of the sheathing. Now run the sheathing diagonally up the gable and nail it on (in this situation, code doesn't require vertical installation). Finally, nail the garage rafters through the sheathing and into the gable wall framing.

Install the rake trim

One of the beauties of building gables flat is being able to do work that would be difficult to do if the gable were upright. I've seen some crews put in the gable air vents and even install siding. But on this house, we just installed the rake trim (that's the trim that fits along the sloping top edges of a gable wall). It's a lot easier to do now, believe me.

Trimming these gables is fairly simple. Start by nailing 1×3 furring strips flush with the top of the rafter and sheathing, as shown in the photo at right.

To make room for the roof sheathing to sit behind the rake board, use a scrap of ½-in. plywood to space the top edge of the rake board above the rafter.

The 1×8 trim meets at the peak in mating plumb cuts. The 1×4 trim extends to the opposite rake and its mating piece is cut to fit against it.

The 1×8 rake boards meet at the peak in mating plumb cuts, but don't do the same thing with the 1×4 trim. Instead, run the 1×4 on one side past the plumb cut on the 1×8s and trim it to match the angle of the opposite rake board before nailing it to the rafter. This process will result in a stronger joint at the peak. The easiest way to mark the board is to hold it in place, trace the angle from below and then flip the board over to make the cut. Mark and cut the intersecting 1×4 in similar fashion. At the eaves end of the gable, let the trim run past the rafter tail for now—you can cut the trim to length later when you install the soffit and fascia.

Raise and brace the gables

The height of a gable wall can make it top heavy and hard to lift, but these gables were small enough to handle with the four-man crew. With a smaller crew, we would have used wall jacks. In any case, this isn't work you should tackle on a windy day or as the last thing on a Friday afternoon before quitting for the weekend. You shouldn't depend on the braces to defend against unexpected winds.

When the gable is nearly vertical, two of us held the wall steady while the other guys nailed 2×4 braces to the studs and to blocks nailed into the floor joists. Double up the blocks so you'll have more wood to nail the braces into. Brace the walls near plumb, but don't try for perfect just yet. Nail the gable plate to the deck at this point with two 16d nails beside every stud.

Straighten and plumb the gables

The gable walls are taller than standard walls, and the rafters are more flexible side-to-side than is the top plate of a wall. Even if you plumb the bottom of the gable, the top can be out of plumb enough to throw off the ridge position and layout. That's why gable walls have to be straightened vertically and standard 8-ft. walls don't.

The best way to straighten a gable wall is to brace it and check it for plumb at several points vertically.

The gables on this house are small enough to be raised with a small crew. Lift the gable in stages: Lift it partway, rest it on sawhorses, and then let everyone change their grip to raise the wall the rest of the way.

Cutting a Scarf Joint

The main rafters on this house were almost 20 ft. long, including the tails. You probably won't find trim boards that long, so you'll have to piece multiple lengths together. Butt joints can open up over time and then they'll stick out like a sore thumb, so I prefer to join all trim boards with a beveled scarf joint. To make the joint, set the bevel angle on your circular saw to 45 degrees, and cut square across the 1×8. Next, cut a matching 45-degree bevel on the end of the next board, slide the two boards together, and nail them on both sides of the joint. If the boards separate, the gap will be less noticeable. To keep water out of the joint, be sure that the higher board on the rake overlaps the lower one.

Diagonal braces nailed to double blocks hold the gable upright until it can be straightened and plumbed.

Dormers are essentially miniature houses, so the framing is usually just a scaled-down version of the house framing.

To straighten and plumb the gable, use 2× stock nailed to the middle and near the top to push and pull the gable wall until the top and bottom are plumb. Then nail braces to blocks on the floor to keep the wall in position.

If the wall is way out of whack, it might take a couple of people pushing and pulling to get it perfect. Unlike the first-floor walls, however, the gable rafters do not have to be straightened along their length at this point. That step is done at the final stages of roof sheathing.

Framing the Dormer Gables

Dormers are a great way to add character to a house exterior. A dormer can create usable floor space without having to raise the entire roof. In fact, the dormers on this house are large enough to require reinforced framing in the floors below. But they were well worth the extra work. The front dormer provides space for a comfortable stairway to the second floor,

and the back dormer is big enough to include a full second-floor bathroom. Both are gable or doghouse dormers—essentially just smaller versions of the main roof. But dormers come in a lot of other shapes, too (see "Dormer Variations" on p. 149).

Like the gable walls of the main roof, dormer gable walls are an interesting mix of wall framing and roof framing. Most of it is the same as for the main roof gable walls so I won't repeat all of that in detail. But the dormer framing does have some different details, and dormers include sidewalls, called cheeks or cheek walls, so I'll focus on those aspects in the rest of this chapter.

Frame the dormer gable walls

The gable walls for the dormers are fairly large (12 ft. wide and almost 11 ft. tall), so it makes sense to build them after the main roof gables are standing. The

Second-Floor Plan

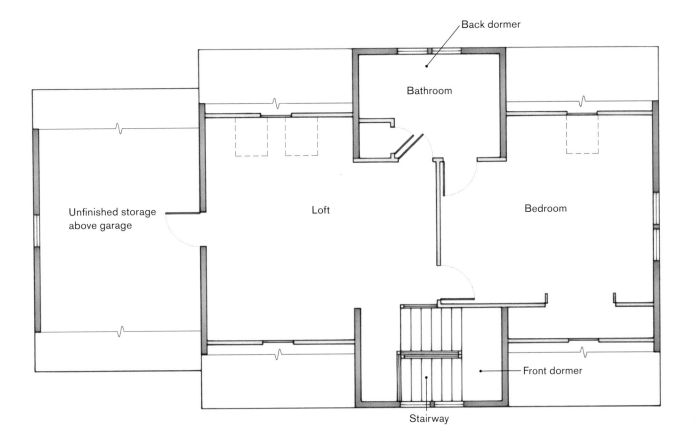

Back dormer

Bathroom

Unfinished storage above garage

Loft

Bedroom

Front dormer

Stairway

cheek walls for the dormers were built and installed at the same time to lock the dormer gables in place.

Install the plates and posts

The exterior framing of both the front and the back dormers was identical. Determine the width of the dormer (12 ft.) by consulting the plans and then cut plates to length. Positioning the plates side-to-side is pretty straightforward because both dormers are centered on the house walls. However, on this house, the front dormer has to be built so that the right-hand cheek wall is flush with the framing of the stair chase (the idea is to create a visually seamless transition between floors). To make sure the right cheek wall will be positioned properly, set the right end of the dormer gable wall plate 5½ in. over from the inside edge of the stair chase framing.

Toenail the bottom plates to the lines snapped earlier when you squared the second-floor deck. You don't have to attach retaining straps to the plates because the dormer gables are a manageable size to lift into place.

The ends of the first-floor walls have L-shaped corner posts (see chapter 4). That's because the walls have top and bottom plates to capture both legs of the L. But because the gable walls of the dormer have no top plate, we opted to make the corners out of tripled 2×6s, creating solid posts under the rafter ends. The length of the posts came directly from the section view page of the floor plans (see "Dormer Framing Section" on p. 148).

The plans set the height from the floor sheathing to the top of the double top plate on the cheek walls at 80 in. There are no top plates on the gable walls, so the tops of the posts had to be the same height as the top

Dormer Framing Section

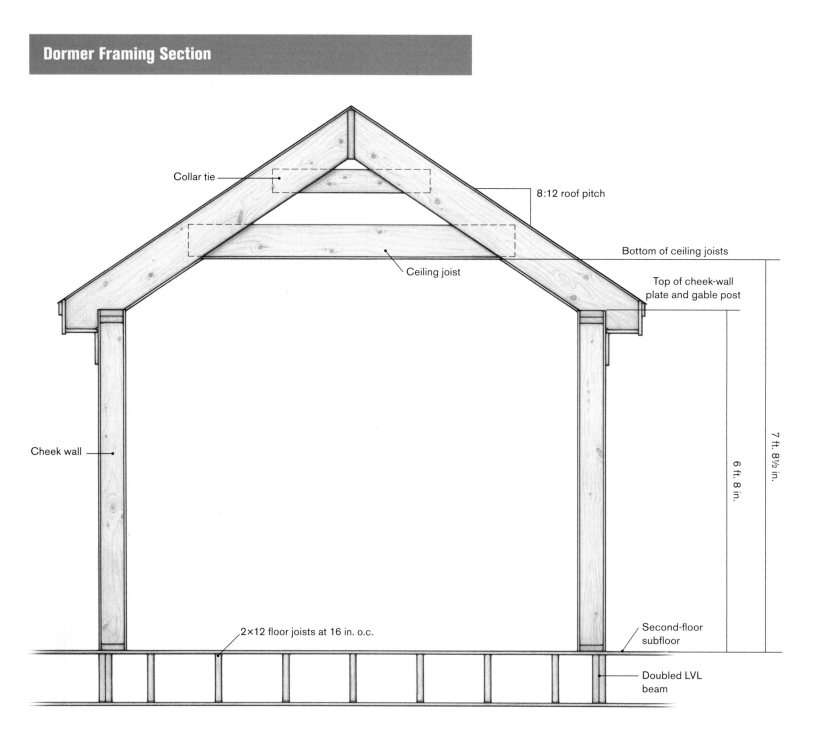

Collar tie

8:12 roof pitch

Ceiling joist

Bottom of ceiling joists

Top of cheek-wall plate and gable post

Cheek wall

7 ft. 8½ in.

6 ft. 8 in.

2×12 floor joists at 16 in. o.c.

Second-floor subfloor

Doubled LVL beam

plates of the cheek walls. (The cheek-wall plates butt into the gable wall instead of overlapping as they did on the first-floor walls.) The posts sat on the bottom plate, so we subtracted 1½ in., resulting in a post length of 78½ in.

Nail the 2×6s together with three 16d nails every 16 in. to assemble the posts, then attach them to the bottom plate by nailing up through the plate and into

their ends with three 16d nails into each 2×6. Next, tack a 12-ft. 2×6 across the tops of the posts to keep the posts parallel to each other (see the top photo on p. 150). The walls are fairly small, so you can square the assembly using a pair of 25-ft. tapes. Tack the posts to the deck on both sides to keep them in place. (To avoid an unpleasant surprise, make a mental note to remove the toenails before you try to lift the gable!)

Shed dormers add the most amount of extra headroom and usable space to the rooms they serve.

Like miniature houses built on the roof, gable dormers create interest on the exterior and let lots of light into the interior.

The steep pitch of an A-dormer roof can contribute a dramatic element to a roof, but these dormers don't add much usable space to the interior.

Dormer shapes can be combined in many ways. Here two gable dormers are joined by a shed dormer.

Dormer Variations

The look of a dormer can vary tremendously depending on the shape of the roof and other features. Here's a quick look at the basic dormer types.

Shed dormers

Shed dormers feature a single roof plane sloped away from the main ridge of the house. The roof has a shallower pitch than the main roof and can start at the very peak of the main roof or down a bit, as here. The front and/or side walls of a shed dormer can be in the same plane as the front and side walls of the house, or the dormer can be an island in the middle of the main roof. Long shed dormers can give a house the look of a saltbox design. Shed dormers create the maximum amount of added headroom in an attic area, but inside they're less interesting than other dormers. The main challenge of building shed dormers involves properly transferring the load of the shed dormer roof to the main roof. Be sure to have an engineer design the proper framing and support.

Gable (doghouse) dormers

Gable dormers look like doghouses that sit on the main roof. They feature two sloped roof planes that meet at a ridge perpendicular to the main roof ridge. The doghouse dormers on the project house are big enough to house a whole room, but on some houses they are too small to add much, if any, usable floor space. The primary purpose of small dormers is to provide natural daylight and ventilation. The gable wall of a doghouse dormer can be either flush with the outside wall of the house below or stepped back from it, as here.

A-dormers

An A-dormer is like a gable dormer without sidewalls. It is sometimes called a cat's-ear dormer because of its tall, pointy shape. A-dormers do not create much usable floor space but they do allow taller windows. The biggest drawback to the style is that the roof pitch is typically much greater than the rest of the roof, which makes for some tricky valley framing that I would not recommend for a novice framer.

Hybrid dormers

Dormer shapes can be mixed and matched to create wonderfully interesting roof patterns. Shingle-style homes are notorious for having unique combinations of dormer shapes. One of the simplest of the hybrids is a shed dormer flanked by two doghouse dormers. I've also seen many shed dormers with an A-dormer in the middle. With all the combinations of intersecting rooflines and wall planes in a hybrid dormer, construction is well beyond the skills of a novice framer.

To ensure that the dormer gable is square and the posts are parallel, check the diagonals and nail a temporary spreader across the tops to secure them. Note that one side of the dormer will be lined up with a wall of the stair chase.

Cut and install the rafters

The rafters for the dormer gables come next. On this project, they have an 8:12 pitch. The total width of the dormer is 12 ft., so the rafter run is 6 ft. Using rafter tables, you'll find that the length of a rafter per foot of run for an 8:12 pitch roof is 14.42 in. So our theoretical length for the dormer rafters was a hair over 86½ in. (14.42 × 6 = 86.52).

Lay out the rafters just as you did for the main gables: Remember to take ¾ in. off the top plumb cut to account for the thickness of the ridge and lay out the rafter tail with the same amount of overhang as on the main roof. On the main rafters, we extended the seat cut ½ in. to accommodate the wall sheathing. Do the same thing with the dormer rafters to account for the cheek-wall sheathing. Make and check a test rafter, and use it as a pattern to mark the others. Then cut the rafters.

The dormer rafters are shorter than those for the main roof so you'll need only single blocks to hold them off the deck. At the bottom of the dormer rafters, we simply toenailed the seat cuts into the tops of the posts to hold the rafters in place during the assembly of the gables. As with the main gables, a ridge block was tacked in place at the peak with a scrap 2×6 holding the rafters together.

Toenail the dormer rafters to the posts (top left). Assemble the peak with a ridge block and a scrap to hold the rafters together (bottom left).

To find the position of the king stud, snap a centerline on the deck, then measure over the proper distance. Use a triangle square to project the measurement to the rafter.

Nail the king studs to the rafters, then frame the opening with jacks and a header.

Build the window openings

Both dormers had identical double windows, and, as with any wall, you can lay out and build the openings first, then fill in the rest of the framing. Installing the wall framing for the front dormer presented an extra challenge because of the stairs, but the process was essentially the same as for the gable walls.

As before, snap a centerline and measure over to lay out the kings for the rough opening. Here's a handy tip: Use a triangle square resting on the deck to project the measurement from the deck up to the rafter, as shown in the top photo above.

Using the same method as with the main gables, measure and install the king studs on both sides of the openings, cutting the flag top on the studs at an 8:12 pitch instead of 10:12.

After the opening is framed, pull the layout for the regular studs from inside the main gable sheathing, and mark the layout on the rafters as well, as shown in the top photo on p. 152. Then measure and install the studs and cripples on their layout points.

After all the studs are cut and installed, slip the ridge post into place above the header and directly below the ridge block, as on the main gables. Framing for the back dormer is a mirror image of the front dormer.

Gable sheathing and trim

Nail on the gable sheathing next. Let the corners of the sheets hang over the top edges of the rafters. When the sheets have been nailed in place, snap a chalkline along the edge of the rafter, and cut off the excess sheathing.

Mark the stud layout for the dormer gable by measuring from the right end of the house (left). The distance from the king stud to the nearest layout should be measured at the plate and then transferred to the rafter (below).

Let the sheathing run past the edge of the rafter, and then nail it off (above). Snap a line along the edge of the rafter, and saw off the excess sheathing (right).

The trim comes next, starting with the furring strip spacers (see the photo above). Instead of a plumb cut for gable rake trim, the crew decided to do overlapping joints. (Once the trim is painted, either type of joint will disappear.) Overlapping joints are easier to execute because the cut line is traced on the trim from underneath. Then the pieces can be flipped over and cut.

Be sure to remember to space the rake trim ½ in. above the top edge of the rafters so it will be flush with the roof sheathing. The 1×4 rake trim overlaps in the opposite direction. One nice thing about the rake trim on the smaller dormer gables is that single lengths of board work fine with no splices needed to join them.

Raise and brace the gables

When the gables are sheathed and trimmed, tip them up into place. (If you didn't remember to pull the tacks holding the posts square, you'll realize your omission pretty quickly.) A crew of four had no trouble raising the walls, though they had to work around the stair chase on the front of the house. Once a wall is lifted, a temporary brace holds it upright. Plumbing the gable is not important at this point. The wall gets plumbed when the cheek walls go on.

Instead of joining the rake boards at the peak with a plumb cut, the boards extend to the rafter on the other side where the angle is traced from below. Then the board can be flipped over and cut to the line.

ANOTHER WAY TO DO IT

Cutting Stepped Layers of Trim

When the dormer gables were built, the rake boards were positioned safely and conveniently over the deck. That meant the rake trim could be cut to length easily while the gable was still lying flat. The problem is that the 1×4 steps down to the 1×8 leaving the saw base without support. So to cut through both layers of trim, set a scrap of 1× on the 1×8 to support the base of the saw as you cut through both layers. The scrap will prevent the saw base from tipping down and binding.

Raising the dormer gable is easy work for a crew of four. The trick is not falling into the stair-chase hole.

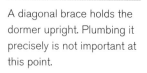

A diagonal brace holds the dormer upright. Plumbing it precisely is not important at this point.

Cheek Walls

Cheek walls sometimes sit on top of the main roof rafters (see "Cheek Wall Framing" below). For this project, however, the weight-carrying framework was built into the second-floor deck in the form of LVL beams. That meant the cheek walls could be built as regular rectangular walls, with the bottom plates sitting directly over the LVLs below.

Frame the walls

With both dormer gables raised, the deck is free for building walls. The cheek walls are the simplest walls in the house so far: just regular studs with a bottom plate and two top plates. Each of the cheeks has an intersecting kneewall that can be built after the roof is framed and sheathed, but instead of installing partition backers now, the crew elected to deal with them when the second-floor walls were being framed. That

Cheek Wall Framing

The sidewalls of a dormer are called cheek walls. They are sometimes built on top of the main roof after it has been sheathed, which makes for some interesting wall framing because the cheeks have to be triangular in shape. In these cases, the rafters are doubled or tripled, and the bottom plate for the cheek wall is nailed through the roof sheathing and into the rafters. The multiple rafters carry most of the weight of the dormer, and the cheeks' shape matches the pitch of the roof. In contrast, the sidewalls for the dormers in the project house actually fit within the main roof framing.

framing detail would not be difficult to add after the fact. No corner studs are needed in the cheeks because the last stud nails directly to the corner posts in the dormer gables.

Start the cheek walls by laying out the top and bottom plates as shown in the photo above. Take the stud layout for the cheek walls from inside the sheathing of the front dormer gable. The total height of the cheek walls is 80 in., so subtract 4½ in. (the thickness of the three plates) for a stud length of 75½ in.

The inside ends of the cheek walls terminate with posts that support the header for the dormer opening in the main roof. The exact positions of the header and those posts are best determined after the main roof rafters are in and the measurements can be marked directly on the rafters, so leave the cheek plates long and cut them to length later. Cheek walls are easy enough to sheathe after they're in place, so there's no need to sheathe them on the deck.

To lay out the plates for a cheek wall, let the tape extend beyond the end of the plate to account for the thickness of the gable wall.

The cheek walls don't weigh much, so they can be carried or slid into place. The first cheek wall to go in is the one that's flush with the stair-chase framing. Nail the plate to the deck and the end stud to the corner post of the gable, leaving the gable's diagonal brace in place. Now pull the nails on the bottom of the brace and plumb the gable. When the gable is plumb, renail the bottom of the brace, and tack it to the cheek-wall studs.

The first cheek wall must be in flush with the stair-chase framing. A diagonal 2× nailed across the cheek wall holds the gable plumb.

To install the other cheek wall, slide it into place and nail it to the gable. Measure over from the first cheek to position the bottom plate of the second cheek, and nail the plate to the deck. As with the other cheek wall, nail on a diagonal brace to plumb that side of the dormer gable.

Sheathe and plumb the walls

The cheek-wall sheathing extends beyond the last stud and attaches to the end of the dormer gable. The outer 4 ft. of the second floor was to be unfinished and unheated space in this house, so we ran a full-width sheet on the first 4 ft. of the cheek walls from bottom plate to top plate. The rest of each wall was sheathed only down to the bottom edge of the rafter; the bottom of the wall would receive drywall.

After cutting a sheet down to 80 in., tap it into place and nail it off. That means you'll have to tap the sheathing into the slot you left earlier at the rafter seat cut. With all that banging on the cheek walls, it makes sense to sheathe the walls before plumbing them side-to-side. Put in the rest of the sheathing after the main rafters are in, but before the main roof is sheathed so that it's easier to access for nailing.

To set the bottom plate for the opposite cheek wall, measure over the width of the dormer.

When the outer portions of the cheek walls are sheathed, plumb them side-to-side. Plumb one wall by running a diagonal brace to a block attached to the deck (see the photo at right). When that wall is braced plumb, measure over from its top plate and brace the second wall plumb.

The installation and plumbing of the cheek walls for the dormer on the back of the house went just about the same. The biggest difference is that the first cheek of the back dormer can be positioned by measuring off the main gable rather than having to align it with the stair framing. That makes locating the wall a snap.

With the dormer braced plumb, nail the first pieces of sheathing into place. You'll add the rest of the sheathing after the main rafters are installed.

To complete the cheek walls, brace one wall perfectly plumb (top). Then measure over at the top plate and brace the second wall plumb (above).

Ridges, Rafters, and Roof Sheathing

With all the gables up, we're well on our way toward getting a roof over our heads. Getting a lid on the house is very exciting because rooflines do so much to give a house its personality. A shallow-pitched roof with wide overhangs has an entirely different look from a steep-pitched roof with diminutive overhangs. And features such as the large dormers on this house add even more character to the house.

In this chapter, we'll complete the roof framing: rafters, ridges, and sheathing. Common rafters essentially lean against each other at the peak of the roof. The ridge ties the rafters together and keeps them spaced properly. And roof sheathing, like wall sheathing, completes the roof and stiffens it into a system that can withstand most anything Mother Nature can dish out.

Installing the Main Roof Rafters and Ridge

STEP BY STEP

Completing the Roof

1   Lay out the rafter plates.

2   Cut and lay out the ridge.

3   Cut the main roof rafters.

4   Install the ridge and the main roof rafters.

5   Sheathe the main roof.

6   Install dormer ridges, rafters, and sheathing.

I once reroofed a two-story ranch house with asphalt shingles. The shingles had already been delivered to the site, and hiring a boom truck to swing the pallets up to the roof was not in the budget. Instead, I carried every bundle up a ladder by hand. After I finished the roof my back was a little sore, but I had a much better appreciation for the real work that roof framing does to carry all that weight day in and day out.

The two main lumber components of roof framing are the ridge and the rafters. The rafters are the hard workers, carrying the weight of the roof system plus whatever weight Mother Nature adds on from time to time. The ridge is like the chairperson of the rafter committee and keeps the rafters together, in order, and in line.

Once the gables are in place, the next phase of roof construction calls for scaffolding. Sure, it's possible to work without it, but scaffolding improves efficiency and safety at the same time, so why *wouldn't* you want it? You can rent pipe scaffolding, but you can also make very strong scaffolding using 2×6 stock to support strong planking (see "Building Scaffolding on Site" on p. 160).

Install and lay out the rafter plates

This house has no sidewalls on the second story, so the lower portion of each rafter rests on a 2×6 plate nailed to the deck. This step was made easier by having the walls of the dormers already built. We just measured the distance between the gable plates and the dormer sidewalls and cut the rafter plates to fit. After the main roof was framed and sheathed, we cut and installed the garage plates to fit between the garage gable and the main house.

After aligning the plates to the lines snapped earlier, nail them flat to the deck with a couple of 16d nails into each floor joist below. The rafters stack directly on top of the joists and studs below at 16 in. o.c., so to mark the layout on the plates, butt the

tape against the inside of the gable sheathing and start the layout from there (see the photo below). Locate any skylights according to the floor plans, and mark the plates for double rafters on either side of their rough openings. Ordinarily, the rafters on either side of the dormers would be doubled as well, but because the dormer on this house was already fully supported by the LVL beams in the second-floor deck, no additional support was necessary.

Install plates along the sidewalls and align them to the snapped lines, then butt your tape against the inside of the gable sheathing and mark the rafter layout.

Building Scaffolding on Site

Sturdy scaffolding can be built relatively quickly on site, using construction lumber already on hand. The key is assembling 2×6 stock into stiff, strong H-frames that can support scaffolding planks, as shown in the left photo below. Use good straight boards, keep the legs parallel to each other, and brace them with a diagonal board. The height of the frame can be anything you want, but we set it so that the bottom edge of the ridge board will be about shoulder high as we stand on the scaffold. That height seems comfortable for working the ridge. The legs of the scaffold should be about 6 ft. apart. Nail a block directly below one side of the horizontal piece for extra support (don't depend solely on nails to keep the crosspiece in place!). The top of the diagonal brace supports the other end of the crosspiece. Position each support about 8 ft. from the nearest gable. Measure from the gable to each leg to ensure that the H-frame is parallel to the gable, then nail diagonal braces to the deck to hold the H-frames upright.

We used 2×12s and plywood to span from each frame to the nearest gable, supporting them on 2× cleats nailed securely to the gable studs. Nail the 2×12s to the cleat and to the H-frames, and then nail ½-in. plywood to the 2×12s to spread out the loads evenly. It's pointless (and dangerous) to build strong H-frames and then load them with flimsy planking, so we used a long, heavy-duty metal staging plank to span between the frames (you should be able to rent one). Set the plank in place and you're ready for the roof framing.

Assembling sturdy H-frames is the first step in building scaffolding. Nail each connection with a half dozen 12d nails. (Note the block that provides extra support on one side of the 2× crosspiece.) A diagonal brace holds the H-frame square and braces hold it steady.

Nail a cleat to the gable at the same height as the horizontal 2× of the H-support frame (top right). Boards nailed to the cleat and the H-support should be covered with plywood. A heavy-duty staging plank completes the assembly (bottom right).

Lay out the ridge sections

The main part of this house is 36 ft. long, so we made the ridge in three pieces. Most of the crews I've seen just make butt joints between the pieces, but we mated them with a 45-degree V-joint. It was a little more work, but you'll see how the effort pays off during installation.

Because the end of the ridge butts against the gable-wall sheathing, the ridge layout is identical to the layout on the plates. All the layout measurements can be transferred directly from the plates. To lay out the ridge, set a nice, straight 2×12 board on sawhorses and mark the rafter layout from one end to the other (see the top photo below). Then flip the ridge over and lay out the other side. Be aware that the rafter layout for the front and the back of the house may not be exactly the same. On this house, the layout for the back rafters changed slightly to accommodate the

A V-joint is better than a butt joint for connecting lengths of ridge board. The joint makes it difficult for the mating ends to slip out of position.

skylights. Be sure to label each ridge piece carefully so that all the layouts on the ridge match up with the plates below.

Making a V-joint Once the ridge board has been marked with the rafter positions from one end to the other, you can lay out the V-joint. Start at the rafter layout position marked farthest from the gable end of the ridge board. Use a large triangular square or framing square to mark a 45-degree angle going through the rafter layout lines. Flip the square, and mark the other side of the joint so that the two layout lines meet at the center of the board. Then cut the V with a circular saw.

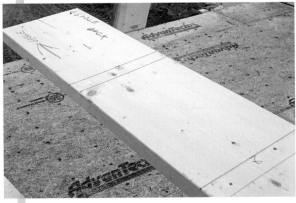

Lay out the first ridge board starting at one end (top). Be sure to label the ends and sides of the ridge piece to keep it oriented properly during installation (above).

To create the mating V-joint, draw a 45-degree angle V through the last layout mark on the section of ridge board. Then start the next ridge section with a complementary V. The rafters on the mark help keep the two ridge sections aligned properly.

Use the pattern rafter to speed layout of the other rafters (above). As each rafter is marked, move it aside and then cut all the rafters as a group (right).

For the next ridge board, mark the rafter positions as you did on the first one. This time, however, start your V-joint layout by drawing a line through the *first* layout position on the board. The idea is to match the ends of each board and end up so that a pair of rafters will bear on each side of the V-joint. When the rafters are nailed on that layout, they align the two sides of the joint and hold the joint together.

Rafter factory

The bundle of 2×10 rafter stock makes a great work station. Two carpenters working together can turn out all the common rafters for a house this size in a remarkably short time.

The first step is to crown the boards in the top layer of the bundle (five boards in our case) and arrange them so that all the crowns face the same direction. Then place the pattern rafter that you made back when you built the gable walls on top of each board, tracing the ends of the pattern onto the board below. To make the process go more quickly, tack a couple of scrap furring strips about 10 in. long to the

edge of the pattern rafter (see the left photo above). Let about 1½ in. of each strip extend below the edge to align the pattern with the board being laid out. The top part of the furring strip makes a handle for easily moving the pattern to the next board.

When all the boards in a layer have been marked out, set the pattern aside and begin cutting. Stack the rafters as you complete them and haul them up to the second-floor deck (this is another time when a fork truck comes in handy!). If you don't have a fork truck, just lean the completed rafters against both sides of the house so they'll be ready to be pulled up and installed.

Ridge and rafter installation

Rafter installation is a task that goes most smoothly if you have some help: a crew of four is ideal, with two at the ridge and one on each side of the house to nail the bottom ends of the rafters to the plates.

Working from the scaffolding, slide the first length of the ridge board into its pocket in the gable wall so that it sits on the ridgepole and against the gable

Special Rafter above the Porch

The roof over the front porch is at a different pitch and will be connected to the lower edge of the main roof later on. The porch rafters are hung from a ledger mounted on the outside wall of the house. If we left tails on the common rafters in this area, they'd interfere with the porch rafters, so we cut nine rafters without tails for that part of the roof.

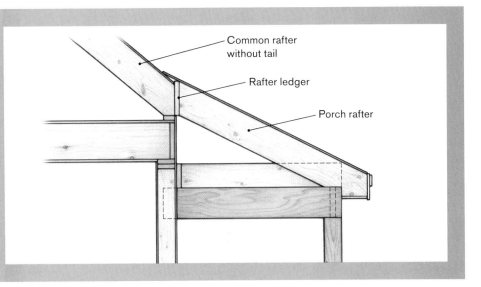

Common rafter without tail

Rafter ledger

Porch rafter

sheathing. Toenail the ridge into the gable rafters. At the other end of the ridge board, nail in a pair of opposing rafters (one front, one back) to hold the ridge up. Then fill in the rest of the rafters on that ridge section, making sure you set the rafters directly opposite each other; otherwise the ridge won't stay in position (see "Setting Rafters" on p. 164).

When the rafters are in place on the first ridge board, position the second ridge board. Nail it through the top edge of the V and into the first ridge board to hold the two together. Then install the rafters, again starting with a pair at the far end to hold the ridge up. Continue in this fashion with the remaining ridge board and rafters. And don't forget that the rafters without tails go on the front of the house over the porch!

A V-joint aligns the ends of individual ridge boards; here's how it looks in place.

Nail one side of the V-joint to the other to hold it together, then place an opposing pair of rafters on the layout marks.

Setting Rafters

There's a trick to setting each rafter into place. First, put it into position and lift it slightly at the ridge so that the bird's mouth at the other end can be snugged up against the wall sheathing. When the bird's mouth is snug and on its layout, toenail the rafter to the plate with two 16d nails on either side (top photo at right). Then put the top end of the rafter on its layout mark and nail through the ridge into the rafter plumb cut with five 16d nails. After the bird's mouth of the opposing rafter is nailed in, set the top of the opposing rafter on its layout mark, and angle five nails in from the back side (bottom photo at right). Nails at the ridge plumb cut are there just to keep the rafters on layout so you have to drive them only along one side. Install the rest of the rafters on the ridge board in the same way.

Frame for the skylights

When you frame a wall, you assemble the openings before you install the regular studs. On a roof it's the opposite. Install all the common rafters to keep the ridge straight, then go back and frame any openings. This house has three skylights on the back side of the house.

On both sides of each skylight bay, the rafters must be doubled. Nail them into place as before, and nail them together along their length as well. The skylight openings are headered off at the top and bottom. The bottoms of the skylight openings line up with the kneewall that will be built on the second floor. So first locate the position of the kneewall from the floor plan and plumb up to the rafter (see the top right photo on the facing page).

Next, measure up and mark the top of the skylight opening, and measure from that point to the top of the rafter. To make the skylights line up perfectly, measure down that same distance at the skylight position farthest away and snap a line between the two points for the top headers. Square down the sides of the doubled rafters from these points for the header layout.

Skylight manufacturers make units that fit conveniently in one or two rafter bays, so the doubled rafters set the width and the headers complete the rough openings. For the length of the headers, measure between the doubled rafters at the bottom plate (the measurement will be most accurate there). Each skylight header is simply two lengths of 2×10 nailed together. To install a header, toenail it to the doubled rafters from the top and sides. Just as with the floor joists and the LVL beams below, joist hangers will be installed later to reinforce the connection. Don't worry: The building inspector will make sure you don't leave out any of those babies.

The short rafters that fill in above and below the skylights are called jack rafters. The length of the jacks is taken by measuring from the top of the rafter to the top header and from the bottom header to the end of the rafter tail. The pattern rafter can be used to trace

Rafters on each side of a skylight should be doubled. Nail them at top, bottom, and into the adjacent common rafter.

Doubled 2×6s serve as skylight headers. Note that in this case they are installed square to the rafters.

Chalkline marks the tops of the skylights.

To position the skylight openings, first plumb up at the kneewall location for the bottoms (top). Locate the top, then measure down that distance from the ridge at both ends. A snapped line across the rafters keeps the skylight framing in line with the ridge (bottom).

Jack rafters

Jack rafters nail in above and below the headers to complete the framing around the dormer openings.

the tail detail. Each jack rafter terminates at the header with a square cut, so preparing them goes quickly. Pull the layout for the jacks by hooking your tape on a nearby common rafter. Toenail the jack rafters to the header; joist hangers will be added later.

Install the posts and dormer headers

A dormer header carries the weight of the main roof around the dormer opening. In this project, the dormer headers are doubled LVLs supported by posts located at the ends of the cheek walls. And what supports the posts? Recall that we incorporated top-loaded beams into the second-floor deck (see p. 107). So far they've been mostly napping, but now it's time for them to take on the load of the headers and the roof above. Build each post from three 2×6s, and spike it to the ends of the cheek wall framing. The posts sit on top of the floor sheathing directly over the beams.

The plans specify that the dormer headers be *flush framed*. In other words, the bottom edges of the headers would be in the same plane as the ceiling joists

Headers will support the roof above the dormers. The headers sit on posts made from tripled 2×6s (top right). After lifting the headers into position (middle right), secure them by nailing through the adjacent rafters (bottom right).

Install jack studs with reverse bird's mouths to fill in the area above the dormer headers.

The ridge for the garage is simple: a single board supported by the gables at either end. A pair of opposing rafters in the middle of the ridge holds it straight.

for the second floor that we'll install along with the second-floor interior walls. So the height of the posts is the same as the height of the second-floor ceiling joists, a figure we took from the plans. The headers are short enough to preassemble and lift into place. Toenail the header to the post, and drive nails through the adjacent rafter and into the ends of the header.

Jack rafters complete the roof framing above the header. These jacks have a plumb cut at the top and a notch cut (called a reverse bird's mouth) at the bottom to wrap around the header. The dimensions and placement of the notch can be taken directly from the rafter that the header attaches to. First measure the length of the rafter to the edge of the header. Then make a seat cut as wide as the header dimension and plumb down from that point. Jack rafter layout can be taken directly from the adjacent common rafters. Toenail the jacks to the header first, then nail them to the ridge.

Garage and remaining details

The garage rafters are all commons, so they go in quickly. The ridge is a single 2×12 board spanning

from the garage gable to the main house gable. Like cutting commons for the main house, we set up a rafter factory using the garage pattern rafter we'd made earlier for the gables.

After setting the ridge between the gables, nail in a pair of opposing rafters near the midpoint. These first rafters keep the ridge in a straight line. Now it's just the simple matter of filling in the rest of the rafters in opposing pairs.

There are a couple of minor framing issues to take care of before the sheathing can go on. On this house, the first is to install solid blocking every 4 ft. between the last two rafter bays of the main house to conform to the wind code requirements.

The last items to go in are the 2×6 collar ties. These framing members supposedly help prevent a roof from

Collar ties join each pair of rafters together just below the ridge.

To keep the courses of sheathing parallel with the ridge, measure down from the top of the rafters and snap a line to guide the first course.

sagging by holding the rafters together, though engineers have arm-wrestled for years over whether this strategy actually works. Still, they were specified on the plans so we put them in. To install ties, snap a line on the underside of the rafters to guide the height as specified on the plans. Nail one end of the collar tie to the line, level the tie, and then nail the other end to the opposite rafter. The plans specified that every collar tie be nailed with five 10d nails at each end.

Installing Roof Sheathing

The roof sheathing for this house was ½-in. plywood. Unlike the deck sheathing, roof sheathing does not have an interlocking tongue-and-groove edge and is not glued to the rafters. To nail the sheathing, use the same ring-shank nails used on the wall sheathing because they have superior holding power. And as with

the wall sheathing, set the nailer so that it does not overdrive the nails. The nail head should just meet the surface of the plywood without breaking the outer ply.

Align the sheathing

Though you can lift the sheathing from the ground up to the second floor at this point, you wouldn't have to if you had anticipated the problem. It's a lot easier to stack the sheathing on the second-floor deck before installing the rafters. That way, one person can cut the sheathing as needed and hand the pieces to the others installing the sheathing. As with the floor sheathing, it's imperative that the roof sheathing be installed in straight courses along snapped lines. On the top edge of the gable rafter, measure up 4 ft. from the lowest point. Then measure down from the top of the rafter to that point (see the photo above). Measuring from the top of the rafter ensures that the sheathing courses are installed parallel to the ridge. Make the same measurement at the rafters beside the dormers, then snap a line between these points to guide the bottom

It normally takes two to set the first course of sheathing. One holds the sheet to a chalked line while the other nails the sheet to the rafters.

ANOTHER WAY TO DO IT

Sheathing Solo

If you find yourself putting on sheathing alone, here's a trick to make the job go more smoothly. Before you position the first course, partially drive nails into the ends of a couple of rafter tails. You can rest the sheathing on the nails as you align the edge of the sheet to the chalked line.

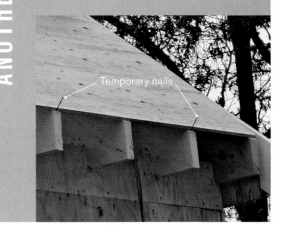

Temporary nails

course of sheathing. This course is the toughest to install because the plywood sheet has to be held on the snapped line and nailed at the same time. It's best done as a two-man job. Don't nail the sheathing to the gable rafters just yet, however. (When all the sheathing is installed, the rakes will be straightened with the sheathing holding them straight.)

Run the courses

When the first course of sheathing is secure, start the next course with a half sheet so that the end seams are staggered. To make installation of the upper courses safer and easier, install toe boards as you work (see "Working on a Roof" on p. 170). Nail the first one about halfway up the first course, and then every 6 ft. or so after that. Drive 16d nails through the 2× toe boards and into the rafters below the sheathing to hold the toe boards securely.

The third course starts with a full sheet like the first course, and it's the last course that butts into the dormer walls. Note that this course covers the skylight

Begin the second course with a half sheet. This approach will prevent vertical seams from overlapping those of the lower course. Toe boards make life on the roof a lot safer and easier.

Working on a Roof

Working on a roof really ratchets up safety concerns on the job site. A sloped plane is challenging enough, but roof sheathing is inherently slippery. Some locales require workers to wear harnesses and be attached to fall-arrest devices when working on a roof, which certainly make the job safer once you get used to them. If you're doing your first roof, I recommend buying or renting the gear. You'll still want to install toe boards, however.

openings. The roofers will cut the openings when the shingles go on and the skylights are ready to be installed. In the meantime, the covered openings will keep out most of the rain.

The fourth course of sheathing runs the length of the house above the dormer openings. Again, measure down from the ridge to keep this course in a straight line. The bottom of the course hangs over the dormer header a bit, but it's okay to trim it later.

The measurement taken from the ridge to align the fourth course also happens to be the width of the top and final course. Before moving to the other side of the dormer, the first two sheets for the top course can be ripped to the proper width, set in place, and nailed.

Let the other end catch up

Now move to the other side of the dormer and sheathe up to meet the courses from the other end of the house. As with that end, measure from the ridge and snap a line to set the first course.

As the courses progress upward, it's important that the third course lines up with its seams staggered below the fourth course coming across from the other side. That meant that the third course had to start at the dormer with a full sheet. The sheets in the fourth course then went in with their seams staggered perfectly.

Straighten the rakes

From there to the ridge, the rest of the sheathing went quickly, finishing off the main house except for nailing the gable rakes. A quick eyeball along the rake tells you just how much tweaking you need to do. When you think of the layers involved, the tops of the gable studs, the gable rafters, and the layers of rake trim, it's little wonder that the rakes can start resembling a winding road before they're straightened. And when walking by a house, there's no sore thumb that stands out more than a crooked rake.

Straightening the rakes is not nearly as complicated a process as wall straightening, but it usually involves two or three people and a chalkline. Near the peak of the roof, drive a nail partway, in between the two layers of rake trim. Hook the chalkline on the nail

The next course runs from gable to gable across the dormer openings. Measure from the peak to position the course. That measurement is the width of the pieces for the top course.

Run the sheathing up the other side of the dormer, making sure that the seams are staggered properly with the course coming across.

To straighten the rake, stretch a string from the peak to the bottom of the rake trim. Move the rake in or out as needed and nail through the sheathing to hold the rake in place.

and then pull it down the rake. As you pull the line tight (no need to snap it), it becomes obvious where the rake trim deviates from the line.

If there are only slight variations, the person nailing off the sheathing should be able to work down the rake, adjusting it in or out until it lines up. Most often a hammer tap is all that's needed to move the rake in, and using a small prybar or twisting the hammer claw against the edge of the plywood moves it out. The weight of the crew member on the roof sheathing usually keeps the rake in place until it can be nailed. In extreme cases, bring in another person to push the rake into alignment while you nail it off.

When the seam between trim layers is perfectly in line with the chalkline, nail through the sheathing into the gable rafter. Keep working down each rake until it's completely straight. Go for perfection on this one; straight rakes make the house look sharp and will be a source of pride for you.

Install the second top plate on the cheek walls before framing the dormer roof. Notice that the H-frame scaffolding supports were nailed to the header to steady them.

Be sure to straighten the gable that the main house shares with the garage before the garage roof sheathing goes on. Nailing the garage sheathing can lock the deviations in place, making them impossible to remove. Sheathing the garage and straightening the garage gable complete the sheathing process.

Framing Dormer Roofs

Once the main roof has been sheathed, it's time to tackle the dormer roofs. After dealing with the fairly massive rafters of the main house (almost 20 ft. long), the 8-ft. dormer rafters will seem like toys. And the 8:12 pitch feels as level as the front lawn after working

on a 10:12 pitch roof. But don't relax or get complacent. There are still a few curves that these simple little roofs can throw you.

The line where two roof planes meet at an inside corner is called a *valley,* such as where the dormer roof meets the main roof. Like ridges, valleys can be structural or nonstructural. A structural valley includes a built-up beam that runs up the valley line from eaves to peak. But because of the extra framing we included in the floors to carry the weight of the dormers, the valleys for these dormers were nonstructural. Nonstructural valleys (we always called them California valleys) are much faster and easier to frame.

Prepping for the roof

Before the framing actually starts, you need to do some prep work. The first thing is to install the second

To prepare for the dormer rafters, cut off the excess roof sheathing on the main roof.

top plate on the cheek walls, as shown in the photo above. When it's in, mark the rafter layout on top. The rafter layout is the same as the stud layout in the wall below.

As with the main roof, work from staging to build the dormer roof. We built H-frames similar to the ones used earlier. To hold them in place, we just nailed the uprights to the header for the dormer opening. A couple of 2×12s were fine for a temporary platform.

When we sheathed the main roof, we let the sheets hang over the header for the dormer opening. To cut the sheathing back to the header, measure up from

both ends and snap a line. Then cut along the line with a circular saw.

Make the ridges

The ridges for the dormer roofs go from the dormer gables to the sheathing on the main roof. Finding the height and length of the ridge is a three-step process. First, measure from the deck to the top of the ridge post. Add the width of the ridge board to this measurement; the total is the height to the top of the ridge. Hook the chalkline on a nail driven partway into the very top of a gable rafter, and stretch the string back to

the main roof. At the dormer header, have someone hold a tape measure at the height of the ridge while another person moves the string up the roof until it is at the right height. That point marks the end of the ridge at the main roof. Measure from this point to the inside surface of the dormer gable sheathing; that's the length of the ridge to its longest point.

To cut the ridge, make a 10:12 pitch cut at one end of the board (to match the pitch of the main roof where the dormer ridge lands). Then hook your tape on the point, mark the ridge length, and make a square cut there. Install the ridge by toenailing it to the tops of the gable rafters. As with the ridge on the main house, these nails just keep it in place until the rafters go on, so four or five 16d nails work just fine. At the other end, a couple of 16d nails through the top edge of the ridge and into the main roof sheathing will keep it from moving side-to-side until the rafters go on. In fact, drive the nails only partway at this point in case the position of the ridge needs to be tweaked a little.

To determine the length of the dormer ridge, measure to the top of the ridge post and add the width of the ridge to the measurement (top left). Stretch a string from the gable to the main roof, and raise the line to the same height (left). Finally, measure from the gable to the mark on the main roof sheathing to determine the length of the dormer ridge (below).

Cut one end of the ridge with a pitch cut to fit against the main roof. Two rafters hold the ridge straight as it is checked for level.

The dormer rafters that meet the main roof do not have rafter tails. Their ends should be flush with the wall sheathing.

Cut and install rafters

Now you're ready for the dormer rafters. Each dormer has four pairs of common rafters. Cut them just as you did the main roof commons. In addition to the commons, three rafter pairs for each dormer should be cut without tails so that they'll land on the cheek-wall plates inside the main roof.

When the ridge is set in place, a pair of opposing rafters near the middle help hold it straight. As with the other rafters, nail the dormer rafters at the seat cut first. Before nailing the tops of the first pair, set a level on the ridge to make sure it's right. If it needs to go up, the end of the ridge can be shimmed slightly. When the ridge is straight and level, nail the tops of the first pair of rafters and install the rest of the commons. Then install the tail-less rafters.

Frame a California valley

With a California valley, the framing member that runs along the valley serves only as a nailer for the

To add the valley board, run a string from the ridge to the point where the dormer rafter intersects the main roof. Measure the top and bottom angles and cut the board to fit.

Make the edge of the valley board flush with the rafters by extending a straightedge from the rafters over the board.

roof sheathing where the dormer roof overlaps the main roof. To find the length and the angles for the valley member, start by snapping a chalkline from the point of the dormer ridge to the point where the main roof intersects with the innermost dormer rafter (see the top photo above).

To determine the top angle of the valley, set a triangular square with one edge against the ridge and the pivot point at the top of the line. The bottom angle should be the complementary angle to the top. Measure the length of the valley board along the chalkline. Because this piece is just a nailer to support the edge of the sheathing, compound angle cuts at the top and bottom are not really necessary, neither is a perfect fit against the ridge and rafter.

If you install the valley board right on the line you snapped, the edge of the board will stick above the plane of the roof. Not good. You have two alternatives:

Shave down the edge of the board to match the roof plane, or simply slide the valley board in until the edge is in the roof plane. Because the sheathing needs just the edge of the board for support, the second choice is easiest and quickest. To make sure the valley board is aligned with the roof plane, extend a straightedge from the rafters to the valley. When the valley board is just touching the straightedge at the top and bottom, nail it in place.

There is one remaining rafter pair to install at the inboard end of the dormer ridge. These stubby little rafters might not seem like they're doing much, but they're needed for proper support of the sheathing. They're a little tricky to cut.

First find the layout point on the valley board by hooking your tape on one of the common rafters and measuring over. That point represents the long point of the stub rafter. The top of the rafter gets an 8:12 plumb cut like the rest of the dormer rafters, but the bottom calls for a compound miter cut. The miter angle (the angle across the face of the board), is the same as the seat cut for an 8:12 pitch rafter. The bevel angle (the angle of the sawblade) is the same as the 10:12 pitch cut for the main roof. Start by making the plumb cut for the top of the rafter. Then mark the length, and draw the miter angle from that point. Set the angle of the saw and make the cut. Because the piece is so small, clamp it to the end of a sawhorse to make the cut. The piece should then go right in as if it grew there.

Dormer roof sheathing

The final step in framing a dormer is sheathing it, which is usually straightforward. Just measure at the top and measure at the bottom, snap a line, and cut the sheet to fit the valley: simple. But a California valley makes the process a little more time-consuming because it calls for an unusually shaped piece of sheathing on each side of the valley.

Stub rafters complete the dormer framing. The top cut is a plumb cut and the bottom cut is a compound miter cut.

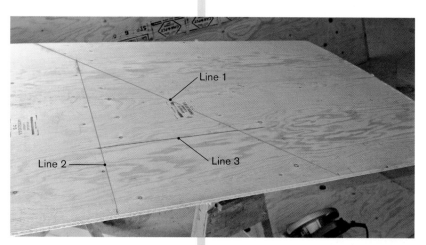

Nail the first piece of dormer sheathing into place. Nails driven partly into the rafter tails help with alignment. Then measure over from the sheathing to determine the size of the odd-shaped piece that will complete the course.

The dormer sheathing tucks into the main roof framing with an odd-looking piece (above). To lay out the sheet, snap a line for the diagonal that represents the valley (1). Square up from the bottom edge for the intersecting line (2) and then draw the last line (3). Cut along your layout lines and fit the piece into place (right). Nail the square tab to the dormer rafters below the plane of the main roof.

Before you get to that piece, however, you have to install the rest of the sheathing. As you did on the main roof, measure up 4 ft. from the bottom of the gable rafter, then measure down from the ridge to that point. This process is especially important with the dormer sheathing because you need to project the line for the top edge of the sheathing into the valley, way beyond a place that you can measure up from. So repeat that measurement at the valley and snap a chalkline between the two points. Begin sheathing at the gable wall. Cut the first piece to length so that it lands in the middle of the last common rafter that has a tail. Align the top of the sheet with the chalkline and nail the leading edge along the centerline of the rafter.

The next piece of sheathing is the tricky one. Measure over to the valley at the top and bottom of the course (see the photo above). Transfer those dimensions to a sheet of sheathing, and snap a chalkline between the two points (see the top photo at left). Next, measure from the edge of the first sheet to the farthest edge of the innermost rafter and transfer that measurement. At that point, square up from the bottom edge of the sheet to the diagonal and snap a second line that intersects the diagonal. Now measure down from the first course layout line to the ends of the tail-less rafters and transfer that dimension to the sheet. (*Note:* Take ¼ in. off as you record and transfer each measurement to make sure the piece fits easily without having to be recut. While a precise fit is nice, it's not really necessary for roof sheathing.) Cut along your layout lines.

Slide the custom piece into place and nail it to the rafters. One last thing: There's a small triangle of main roof sheathing to install. First nail in small 2× scraps as nailers and then drop the last triangle into place. Be sure that this last piece of sheathing stays in the same plane as the rest of the main roof.

Like the other top courses, rip the sheets for the last course to fit. This time, start with a full-length sheet to overlap the end of the sheet on the first course. Filling in this course is much easier. Just measure over at the top and bottom and cut a diagonal for the valley. Drop that piece into place, and the dormer framing is complete.

A small triangular piece fills in the rest of the main sheathing. It attaches to nailers along the header and the rafter. Make sure the piece stays in plane with the main roof.

The last piece of dormer sheathing has a simple shape and will easily drop into place. Be sure that the sheathing is nailed securely before you stand on it.

Framing Interior Walls and Ceilings

With a roof now over your head, you can turn your attention to the interior walls. This house has finished first and second floors. As on many houses, the first-floor walls are pretty straightforward because the ceiling is flat and doesn't change height. On the second floor, however, some of the walls have to fit under flat ceilings and others tuck in against the slope of the roof. Heights vary, too, so you'll have to pay attention when you start work upstairs.

Prep Work for the Framing

Remember the rainy day job of installing joist hangers that you put off a while back? If that day never arrived, put them in now whether it's raining or not. After that, another bit of prep work is to strap the ceilings, a job that may strike some readers as unnecessary (though I don't agree). In any case, both of these jobs must be done before interior walls, called partitions, can be built.

STEP BY STEP

Building Interior Walls

1   Complete prep work.
2   Install ceiling strapping on first floor.
3   Install ceiling joists, then strapping on second floor.
4   Snap lines for wall layout floor by floor.
5   Build and install first-floor partition walls.
6   Build and install second-floor partition walls.
7   Build and install kneewalls.
8   Check all walls and ceilings; install drywall backers as needed.

Joist hangers

When joist hangers first appeared on job sites, carpenters used just about any kind of nail to install them. But often heads would pop off the nails, or the nail shaft would shear off, causing hanger failure. Joist hanger nails were devised in response to those problems. They have a stronger head-to-shaft connection, and the shaft itself is a larger diameter than standard nails to resist shear forces. They're typically galvanized but are also available in stainless steel for exterior use. Joist hanger nails for positive placement pneumatic nailers come in 1½-in. and 2½-in. lengths. Use the shorter length if you're driving the nails into a single thickness of 2×.

Nailing joist hangers into place by hand can be frustrating, in part because you have to work in spaces no wider than 14½ in. (the distance between joists and studs spaced 16-in. o.c.). Also, errant hammer blows inevitably seem to find fingers holding those stubby nails. It's no wonder that installing joist hangers was a job left to the lowest guy on the framing crew totem pole.

Eventually, some brilliant inventor came up with a positive placement pneumatic nailer. With these nailers, the point of the nail sticks out of the nailer slightly to aid in placing it in the joist hanger hole. These nailers make install-ing joist hangers a snap. By the way, there are

two small nailing tabs near the top of a joist hanger. When driven into the wood with a hammer, they hold the hanger in place until you can drive the nails.

Joist hangers *must* be sized to suit the framing members that you're hanging (see chapter 1). For

Use hangers designed specifically for the size and type of material you're hanging. Here, a hanger for a double 12-in. LVL is being tapped into place.

Installing Joist Hangers

To install a joist hanger, slip it into place over the end of the joist. Make sure that the bottom of the hanger is flat against the bottom of the joist, otherwise it will interfere with the installation of strapping or drywall. Snug one side of the hanger against the side of the joist and nail it off. Push the other side of the hanger into position and nail it. Always drive a nail into every hole on both sides of the flange. To save time, install as many hangers as you can from one ladder position. I can usually install three or four before I have to move the ladder.

The holes in joist hangers are there for a reason. Unless every hole is used, the hanger won't develop full strength. In fact, the building code requires a nail in every hole in every hanger.

example, the floor joists of this house are single 2×12s, so the hangers were for single 2×12s. But the beam intersection needed a hanger for double 12-in. LVLs. Make sure you use the proper hanger. LVLs are wider than 2× stock, so LVL hangers shouldn't be used to support dimensional lumber.

First-floor ceiling strapping

After working on houses in southern New England for 15 years, I assumed (incorrectly) that home-building techniques were pretty much the same all over the country. My error in thinking was never so apparent as with the issue of strapping ceilings. I was shocked to learn that builders in most parts of the country had never even heard of the technique. In fact, most thought it was absurd. Larry Haun, a West Coast carpenter and good friend whose opinions on carpentry I value tremendously, put it this way: "Strapping ceilings is a waste of perfectly good 1×3s." Well, those of us who learned to frame houses with strapped ceilings have equally strong opinions, so let me explain why.

Ceiling strapping consists of 1×3 furring strips applied perpendicular to the joists 16 in. o.c. The strips "fair" the ceiling, smoothing out any irregularities in the joists to make the ceiling flatter. With an actual dimension of ¾ in. thick and 2½ in. wide, furring strips offer a much wider fastening surface than the edge of a joist, and that extra width makes installation of interior partitions and drywall go much more quickly.

The furring strips offer support for systems such as ductwork that run through the joist bays. They also provide a space for securing electrical wires *below* the joists so electricians don't have to drill *through* them. Some local officials might take exception to running wire beside furring strips, especially if you're in an area where strapping ceilings is not commonplace. Before you use this strategy, I'd check with your building officials to make sure they're on board with the practice. Around here in southern New England, plumbers and electricians charge more if they have to work on a house that doesn't have strapped ceilings. Strapping is easy to install and usually rather inexpensive.

Installing strapping Like most other framing tasks, installation of the furring strips starts with a layout. Mark the layout either along the top plates or on the nailers you attached to the exterior wall plates (see the bottom right photo on p. 114). The layout for the furring strips is 16 in. o.c., but instead of subtracting ¾ in. at each layout mark as you did to space the joists, studs, and rafters, subtract 1¼ in. (half the width of a 1×3).

After marking the layout at each end of an open space, stretch a chalkline between the walls at each layout mark and snap lines onto the edges of the joists. You can usually snap four of five lines before having to

rechalk the line. The lines space the strips and help keep them straight during installation. Where the expanse of the ceiling is interrupted by walls, such as for the stair chase, mark the layout and snap lines on the joists on both sides of the chase.

Around here, furring strips come in bundles of 10 and can be up to 16 ft. long. The bundles make it easy to cut as many as six pieces to length at one time. Align the ends by tapping them with a hammer. Then mark the length you need on the top strips and cut through the pile. With the blade set at maximum depth, a 7¼-in. circular saw can cut through three layers and score the fourth layer in one pass.

Ceiling strapping is an easy way to flatten a ceiling while creating better nailing for the interior walls and the drywall. (The interior walls shown here are structural walls installed earlier.)

Where long lengths of strapping are necessary, stagger the butt joints for a stronger ceiling.

interior structural wall and the stair-chase wall, nail a furring strip on both sides of the top plate. With one person cutting and another nailing, installation goes quickly if you work across the room nailing several strips at a time. Otherwise you'll waste a lot of time dancing with your ladder.

Second-floor ceilings

On this house, strapping the first-floor ceiling is easy because it's flat. Upstairs is another story. Portions of that ceiling are sloped, and the way the dormers intersect the main floor area makes it a little tricky to finish the framing so that all the edges of the drywall will be supported adequately. Then there's the matter of the ceiling joists.

Ceiling joists There's an old saying: "One person's ceiling is another one's floor." That accurately describes the basement ceiling and the first-floor ceiling, but not the second-floor ceiling. (By the way, the basement ceiling gets strapping only if the basement will be finished living space.) The second-floor ceiling joists are 2× members that span between the rafters on the main roof as well as the dormers. Their main function is to provide a solid surface for attaching the ceiling drywall. Because these ceiling joists do not carry any loads other than the drywall and insulation, they can be smaller than the floor joists. But they still have to span over 10 ft. in places, so the plans specified 2×8s. The plans also specified the height to the underside of the ceiling joists as 92½ in.

To install the ceiling joists, measure up that distance from the floor to the rafters at either end of a section of joists and on the front and back sides of the house. The dormer headers make matters easy because they're already set at the right height. Next, snap lines between the points on the underside of the rafters. On one side, nail a furring strip to the rafters below that line. This furring strip becomes part of the ceiling strapping, so it can be nailed in permanently.

The joists don't have to be cut to an exact length, as long as the bottom edge extends past the edge of the rafter by a few inches at each end. To put the joists in, simply rest one end of the joist on top of the furring

The areas on either side of the stair chase are about 12 ft. long, so a single furring strip can go from one side to the other. For the areas that reach across the whole length of the house, several pieces must be placed end to end. To avoid a weak point in the ceiling, stagger the ends of the strips by at least one joist bay.

To nail the strips, drive two 8d ring-shank nails at each joist (these are the same nails we used to attach the sheathing; the ring shank gives them superior holding power). For every interior partition already installed that runs parallel to the strapping, such as the

To install ceiling joists between rafters as in a dormer, nail a furring strip at the proper height to hold one end of the joist while you nail the free end to a snapped line. (For the sake of safety, the crew nailed 2× barriers across window openings.)

Furring strip

To strap an attic ceiling, put furring strips on all the ceiling planes, starting at the joint between two adjacent ceiling planes.

strip, slide the other end up the joist to the line, and nail it with four or five 16d nails. When the ends of the joists are nailed on one side, go back and nail off the other side that's resting on the furring strip.

At the gable ends, attach a 2×4 nailer to the gable walls at the ceiling joist height to catch the ends of the furring strips. In the middle of the main roof, the ceiling joists butt into the dormer headers, so they have to be cut to fit. Toenail the joists to the headers to hold them in place, then secure them with joist hangers to make the connection permanent. The room over the garage was unfinished and uninsulated, so no ceiling joists were needed in there.

Ceiling strapping
The second-floor ceiling strapping can be installed much as the first floor, with a couple of exceptions. You already nailed in a furring strip to hold one end of the ceiling joists as you installed them, so match it with a strip on the other side of the rafters. These strips serve as the drywall nailer for the top of the sloped portion of the ceiling. Next, fill out the horizontal portion of the ceiling with additional furring strips spaced 16-in. o. c.

Another place the strapping technique differs slightly is at the skylights. After snapping the lines in those areas, we ran furring strips right through the

openings. After filling in around the skylight with short pieces of furring to fully support the drywall edges, we cut the strips from the opening.

On every project, there are areas where special details have to be worked out to support the strapping. One such area on this house was where the posts hold up the dormer headers. The ceiling changed from

At skylights, run the strips over the opening and cut them out later (this saves time). Use small pieces to fill in around the edge of the opening; the drywall contractor will thank you for providing this support.

Add blocks as necessary to support the ends of a furring strip where there is no joist.

sloped to flat at that point, but a full ceiling joist was not necessary because the rafters next to the header fell in between the layout. Instead of wasting a full 2×8, we nailed an angled block to the ends of the headers to provide nailing for the furring strips on both sides of the post.

First-Floor Walls

At this point, three interior walls are already in place: the structural wall that carries the LVL beams in the second-floor deck (see chapter 5), and the two walls that border the open sides of the stair chase (see chapter 6). But now that the ceilings have been strapped, it's time to carve up the rest of the interior with partition walls.

Compared to building exterior walls, building partitions is fast and easy. To begin with, they're made of 2×4s, which are lighter and easier to handle than 2×6s. Interior walls don't have structural sheathing, so they can be squared and plumbed in place. They have no second top plate. And door openings in a nonstructural interior wall can be framed with nonstructural headers, which are much easier to build.

Layout Strategy for Interior Walls

To build the interior walls of this house, the head of the framing crew devised a logical and efficient strategy. Another person, however, might have taken a different approach. And no doubt the layout of walls in your house will be different from that of this project. The important thing is to think through the installation even before you bring materials inside. After all, *you* are now the head of the framing crew.

Begin by looking at the floor plans and the floor space. Think about where you'll stack material and what open areas you'll need for building the walls. Try to avoid having to move that pile of 2×4s because it ended up in the only open space you have left to build the last few walls.

Here's the good news: Without sheathing or wallboard, interior walls are fairly easy to alter. You can add a partition backer or a corner nailer after the fact if your sequence goes a bit awry. You might find it helpful to number the walls on the plans to work out a logical installation sequence. And you don't have to nail home every wall as you build it—you can tack the walls close to their final placement until adjacent walls have been set in position.

First-Floor Plan: Interior Wall

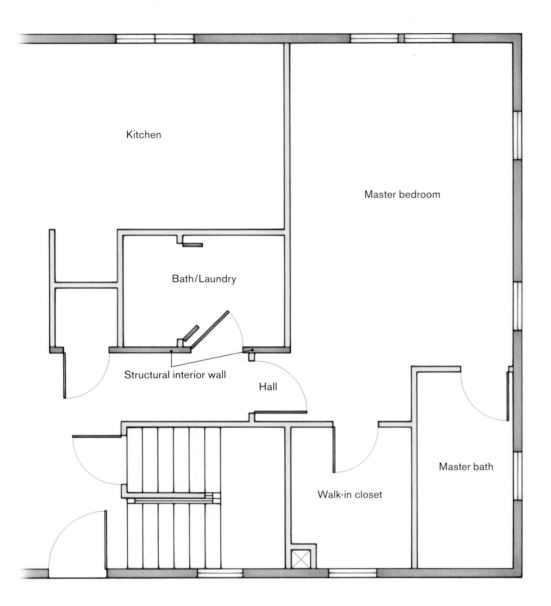

Kitchen

Master bedroom

Bath/Laundry

Structural interior wall

Hall

Master bath

Walk-in closet

But that said, pay close attention to the layout so rooms will be square and the right size. The position of many of the interior walls is indicated by the partition backers in the exterior walls, but it's a good idea to review the floor plans and double-check the backer locations before you start framing. If a partition backer is off layout by more than 1 in., make sure that changing the dimension of the room won't make it impossible to install such things as bathtubs and appliances. If so, you'll just have to reposition the backer.

Measure and snap lines

The trick to locating interior walls is to start at a corner of the house. Working off two perpendicular walls makes it easy to lay out the first partition so it will be square or parallel to those walls. And once you establish one partition wall that's square, you can find the location of all the walls that branch out from it. In this house, we started our layout with the master bath and its partition backers. A partition backer, as you'll recall, consists of two studs along with a nailer that's the same width as the partition wall.

Wall Layout Sequence

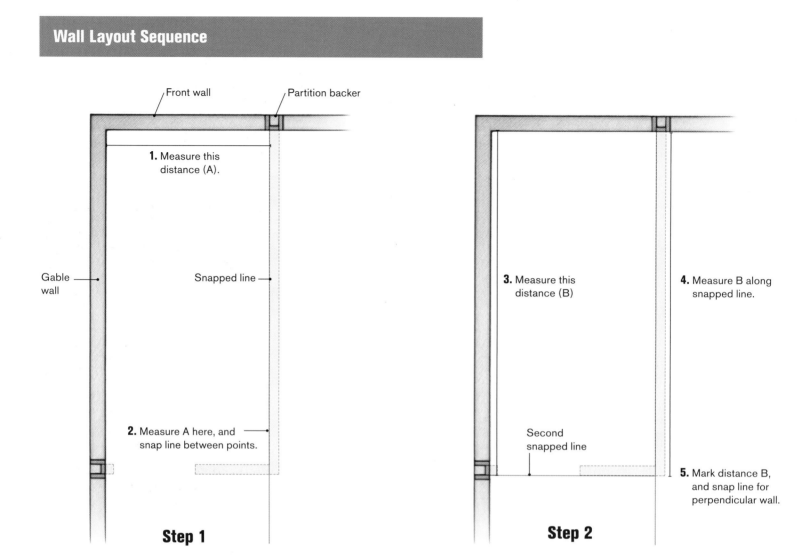

Front wall   Partition backer

Gable wall

1. Measure this distance (A).

Snapped line

2. Measure A here, and snap line between points.

Step 1

3. Measure this distance (B)

4. Measure B along snapped line.

Second snapped line

5. Mark distance B, and snap line for perpendicular wall.

Step 2

Measure over to the near edge of the nailer in the partition backer on the front wall, distance A on "Wall Layout Sequence" above. Now measure out from the backer on the gable wall and mark the same distance on the deck (see the top photo on the facing page). Snap a line from the front wall to the point you just marked, and you'll have a line parallel to the gable wall.

Next, measure from the corner to the far edge of the backer on the gable wall (distance B) and mark that distance on the line you just snapped. Snap a line from the gable backer to that mark. Now you have the length and location of both master-bath walls. Just be sure to put an X on the side of the line where the plates will sit or else the wall will be 3½ in. off layout.

The rest of the partition walls go in pretty much the same fashion. The bathroom wall parallel to the gable is also parallel to the stair chase. So for the walk-in closet wall, just measure out the same distance on the bathroom wall and the stair-chase wall, then snap another line. Now you've completed the layout for all the walls for that section of the house. You can ignore the pair of tiny walls at the back corner of the closet—that's a chase for mechanicals and will be built after any piping or ducts have been installed.

The next wall to locate is the one that separates the kitchen and bath/laundry from the master bedroom. This wall forms a corner with the interior structural wall and lets you once again measure parallel distances in each direction to locate the next series of walls. This line is easy because you don't even have to measure for it. Just stretch the chalkline from the partition backer on the back wall, past the end of the structural wall and to the stair-chase corner. (The wall ends at the

Start laying out the interior walls with a measurement to the partition backer on the exterior wall. Then make the same measurement near where the wall will end (shown here). Snap a line between those points (as shown in "Wall Layout Sequence" on the facing page).

To avoid confusion, clearly mark the side of the line where the plate will sit with an X.

To locate the kitchen wall, snap a line from the partition backer in the back wall to past the end of the structural interior wall.

structural wall, but extending the line past the structural wall confirms that the two walls line up properly.)

The rest of the first-floor interior walls are straightforward, so continue in this manner taking measurements either from the plans or from the backers in the walls already built. Then snap intersecting lines to lay out the remaining walls. Again, be sure to mark an X on the sides of all the lines where the plates will go.

Thinking ahead

The structural interior wall in this house was built like an exterior wall because it supports LVL beams (see chapter 4). But none of the other interior walls are structural (which is why they are often called non-bearing partitions). That means they're a lot easier to frame because the headers are simply 2×4s nailed horizontally to the tops of the jacks (see "Anatomy of a Nonstructural Interior Partition" on p. 190). In fact, you probably wouldn't even need to double the headers, but doing that makes the finish guy's job a lot easier down the line because it provides extra nailing surface for the trim.

Overall, I try to build and raise the walls in about the same progression as I did the layout. Building them in order just helps me keep everything straight.

Anatomy of a Nonstructural Interior Partition

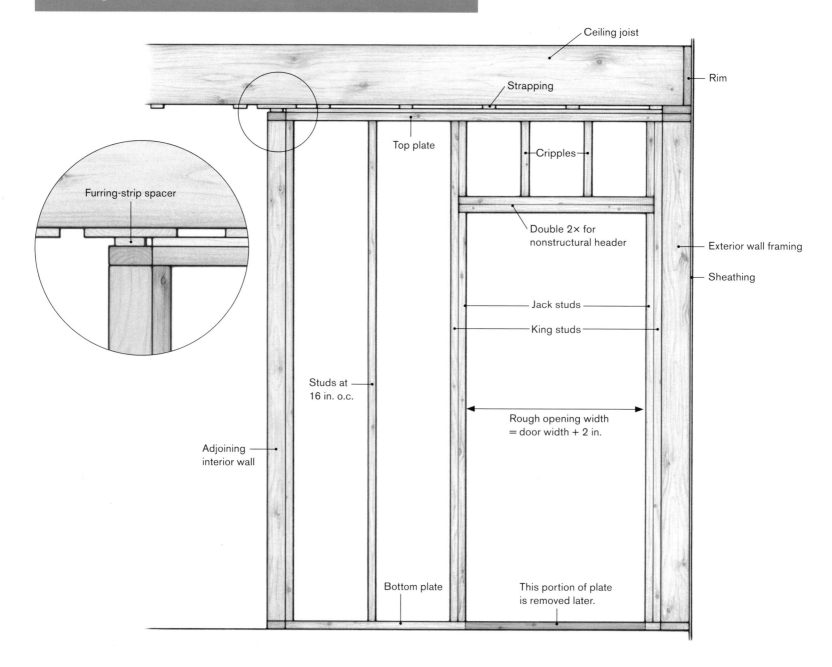

I also pay attention to how walls will be attached to each other. For example, the side wall of the master bath attaches to the door wall of the bath, so the door wall should be built and installed first.

Lay out the plates and rough openings

As with exterior walls, the first-floor partitions start with a pair of plates, but measurement can be taken directly from the layout lines you just snapped. As you

cut the plates for a wall that has one or more door openings, ignore the openings for now and just cut the top and bottom plates as if the doors weren't there. Once the wall is installed, you can cut and remove the doorway plate.

After cutting plates for any partition with a door, check the plans and the door schedule to get the door size and location. The rough opening width for interior doors, including bifold doors, is always 2 in. wider than the door itself to provide enough room to

Walls That Intersect

You'll often hear framers refer to walls as butt walls and by walls. These terms refer to the way in which one wall intersects another. On the exterior walls, the front and back walls are by walls: The ends of these walls extend past the gable walls. The exterior gable walls terminate against the front and back walls so they are considered butt walls. Whether a wall is a butt wall or a by wall isn't a big deal, but it affects layout because it affects the actual length of a framed wall.

The terms *butt* and *by* refer to just the ends of the walls, however, so it is quite possible for a wall to butt against another at one end but extend past a wall at the other end. A good example of this is the back wall of the garage. This wall butts into the gable wall of the house, but runs past the garage gable. With interior walls, figure out which ends will be butt ends and which will be by ends before you start. It will make the task of building those walls go much more quickly.

<h2>Butt Walls Versus By Walls</h2>

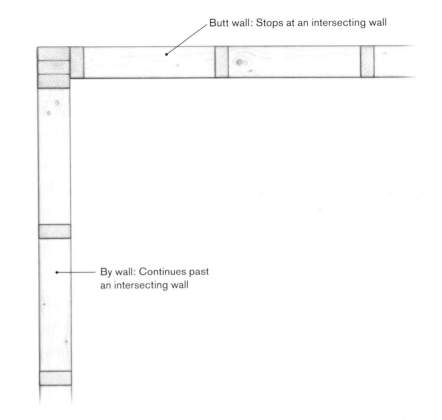

Butt wall: Stops at an intersecting wall

By wall: Continues past an intersecting wall

position and shim the door jamb later. For example, on this house the master bath's 30-in. door requires a rough opening 32 in. wide.

A door's rough opening height takes a little more figuring. Typically, interior doors are installed after the finish floors have been installed (the exception is carpeting, which is installed after the doors are in place). Finish floors vary in thickness, so you have to size the height of the rough opening to make sure there's enough clearance for the door to swing (believe me, correctly sizing the rough opening now is a lot easier than trimming the door later). Typically all the doors on a floor will be the same height so that the header trim lines up when the finish work is done. Just double-check to make sure the height doesn't change for a door in a specific situation.

As a rule of thumb, I just add 2 in. to the height of the door for the height of the rough opening. The extra

ANOTHER WAY TO DO IT

Layout Trick for Tight Quarters

Space was tight in the master bathroom of this house. To leave enough room for the vanity cabinet and countertop, the door had to be as close to the exterior wall as possible. To solve the problem, we framed the wall so that the end stud would also be the king stud for the opening. To do this sort of layout, just start at the end of the plate, mark the king and the jack, measure over the rough opening width, and mark the other king and jack.

space should be plenty for the door to clear just about any kind of finish floor, plus a little for any tweaking of the door height that may be needed.

Just as with the exterior doors and windows, the jacks set the height of the rough openings. For the length of the jacks, start with the overall height of the rough opening. Then subtract 1½ in., the thickness of the bottom plate that will be removed later. So the length of the jacks for an 80-in. door will be 80½ in. (80 + 2 − 1½ = 80½).

Once you determine the rough opening for a wall containing a door, you can lay out the plates. First, lay out the king studs and jacks for the rough opening, then lay out the studs and cripples above the door headers.

Assemble the walls

Interior walls are light, so you don't have to build them on their layout as you did the exterior walls. Instead, build them in the most wide-open space then slide them over to position. With the plates cut and laid out and all the other pieces cut, the wall goes together quickly. As with the exterior walls, attach the jacks to

the king studs first. Nail them in place between the plates, and add the double header. Then go back and fill in the studs and cripples. The studs for interior walls on the first floor are a full 8 ft. Don't bother to square the wall before raising it—it will square itself once you nail it to the plumb exterior wall.

Where walls run parallel to the strapping, fasten an additional ¾-in. spacer to the joists to provide a nailing surface for the drywall on either side of the wall. The spacer also serves as a nailer to attach the top wall plate where there isn't any strapping (see the right photo below). Enter our old friends, the leftover springboards. You should still have plenty in your pile. Their 7½-in. width is perfect for providing drywall nailing on both sides of the 2×4 top plate. The length of these boards is not crucial, but make them long enough to catch the joist that is just beyond the end of the wall.

Before raising the first bathroom wall, we cut the plates for the next bathroom wall and set them in place to mark the stud layout. The partition backers are made the same way as for exterior walls. To locate them on the plates, just measure to any intersecting wall line and put a partition backer where the walls meet.

Build partition walls flat on the deck one-by-one, and then lift them into place.

Use old springboards as nailers for walls that land between furring strips. Locate and attach these nailers after the wall layout has been finalized.

Topping the Wall with a Furring Strip

When building the interior walls, the crew working on this house used an interesting technique that I'd never seen before. They added a 1×3 furring strip on top of each wall's top plate before tipping the wall into place. The extra furring strip makes sense for a number of reasons. First, combined with the ceiling strapping, the top plates of the partitions would be at the same height as the lower top plate of the exterior walls. That made the interior studs the same length as the exterior studs, or 8 ft., for convenient framing without a lot of cutting. This technique also provides better nailing for the drywall. The added furring strip leaves a gap at the top edge of the wall, but with the ceiling drywall installed first, that gap is only ¼ in. or less. And that leaves a full 1½ in. of top plate left to attach the top edge of the drywall sheets on the wall.

The final reason for adding the furring strip to the top plate is more subtle. Building interior walls flat and then lifting them into a vertical position can be frustrating because the diagonal length from the corner of the top plate to the opposite corner of the bottom plate is longer than the wall is high, which means the walls have to be forced into place. To avoid this, I've known builders who frame interior walls short and then shim between the top plate and the strapping. But with a furring strip roughly centered on the top plate, that diagonal distance is much closer to the overall height of the wall, and the wall is much easier to raise into position with a minimum amount of effort.

Plate length and layout can be taken right from the layout lines on the floor. Laying out the studs is easier with the plates in place. The walls are then built in a wide-open area of the floor.

A furring strip on top of the wall makes it easier to tip the wall into position (above). After nailing the plate to the snapped line and plumbing the wall at the backer, the intersecting wall is tapped into position (left).

When the wall is up, nail the plate to the deck.

Install the walls

With the first two walls built, it's time to get them into place. Set the first wall so the bottom plate is close to the layout line, then push it up and into position. When the wall is vertical, tap the bottom plate until it is aligned with the snapped line, and then nail the plate to the deck with two 16d nails at the base of every stud. Plumb the end of the wall that meets the exterior wall and nail it to the partition backer with a couple of 16d nails every 16 in. or so top to bottom. Leave the top plate free for now. Next, bring the second wall over and position it. You may need to tap it into position with a sledge, but the furring strip on top should give you enough wiggle room to avoid having to bash the wall into place.

Snug the bottom plate to the adjoining walls, align the plate to the layout line, and nail it to the deck. Plumb the end of the wall that meets the front wall of the house, and nail the last stud to the partition backer. Then snug the two interior walls together at the corner and make sure they're plumb in both directions. If the plates were cut accurately and the exterior walls are plumb, the corner should be perfect. If it's not, tap the offending wall in or out at the top or bottom until the corner is plumb in both directions. Finally, nail the top plates to the furring strips (or the 1×8s, depending on the wall) at every joist or every 16 in., leaving plenty of room for plumbers and electricians to drill through the plates without hitting nails. With the bathroom walls up and secure, measure from the bathroom wall to the stair-chase wall to determine the length of the walk-in closet wall plates and install that wall as you did the first two. Leave a good ⅛ in. of play in the plate length so the wall will slide in easily.

Installing a Long Wall

Long partition walls can be tricky because their size often restricts your ability to maneuver them into position. The longest partition in this house is the one that separates the kitchen from the master bedroom wall, so we built it with the bottom plate against the layout line. After it was framed, we just had to lift it. After nailing the bottom plate and the end studs into position, we plumbed the partition backer before nailing the top plate to the ceiling. That ensured that the intersecting partition would be square. Always check a long wall at several studs to make sure it's plumb.

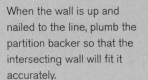

When the wall is up and nailed to the line, plumb the partition backer so that the intersecting wall will fit it accurately.

Long walls are awkward to move once assembled, so build them close to their final position.

Once you understand how to install these walls, continue to build walls in a logical sequence as you work through the house. About the only challenge you'll encounter is building and installing unusually long walls (see "Installing a Long Wall" above).

Remember to put the walls up in the correct order. That little wall at the back of the front hall closet, for example, will be easier to install before the left-hand wall and back wall of the laundry room are up. Small walls such as the door wall into the master bedroom are easy to build and carry over to the correct position. Walls for closets with bifold or sliding doors don't typically need structural headers—nonstructural headers will do just fine because the weight of the doors is not significant.

Short walls are portable. Build them where you have room, and carry them to where they will be installed.

Time-Savers for Layout

To save time (and improve accuracy) when laying out walls, look for opportunities to steal a layout you've already done. For example, when laying out the plates for the laundry wall opposite the long kitchen wall, we slid the plates against the kitchen wall and transferred the layout directly. Likewise, a quick way to lay out the plates for small walls is to hold them against the related section of wall above the header so you can transfer the position of the cripples.

To speed the stud layout, set the plates against the opposite wall and transfer the stud locations directly to the plates (above). For short walls, cut the plates and hold them against the cripples to transfer the stud locations (left).

Second-Floor Walls

Now for the upstairs! On this house, the second-floor walls are shorter because the ceiling is only 7 ft. 8½ in. high. And with the rafters squeezing the floor space from the sides, quarters are a little cramped. When space is tight, it's particularly important to think about which walls to build first.

Lay out the main walls

There are really just two main walls on the second floor that separate the general spaces (see "Second-Floor Plan" on p. 147). The bathroom wall sits between the posts that support the dormer header, so start by snapping a line along the inside edges of the header posts.

The other main wall separates the upstairs bedroom from the loft area. Get the location from the plans, measure from the gable end, and snap a line from the bathroom wall to the stairway for that wall.

Upstairs, snap a line between the dormer header posts for the bathroom wall (left). Then measure over and snap a line for the wall separating the bedroom from the loft wall (below).

Locate and snap lines for the linen closet and the bedroom closet walls just as you did for the smaller walls on the first floor.

Build the main walls

When quarters are cramped, it's crucial to build walls in the right order. We started with the bathroom wall because the separation wall butts up to it. Besides, the separation wall would be in the way if we installed it first.

Measure between the posts to get the length of the bottom plate of the bathroom wall (see the top photo on p. 198). But the top of the wall runs between sloped sections of the ceiling, so the top plate has to be shorter than the bottom plate. To complicate matters, this wall includes two door openings, and they're *not* centered on the wall. All the more reason to pay close attention to the measurements on the plan. Lay out the bottom plate first. Then measure the flat part of the ceiling and cut the top plate to that length. Position the

Adding Studs

Framing partition walls isn't difficult, but it's easy to forget a stud here and there, particularly when you're new at this. So don't worry if you find you have to add framing. When adding a stud or backer after a wall is in place, toenail it to the top and bottom plates, using two 16d nails on each side of the stud.

top plate on the bottom plate and transfer the layout (the wall is centered in the space, so the top plate is centered on the bottom plate).

Build these door openings with nonstructural headers. Leave the "extra" length of the bottom plate at either end of the bathroom wall free at this point without end studs. After building the wall, raise it, tap it to the line, and nail the bottom plate to the deck. Next, plumb the studs at either end of the wall before nailing the top plate to the furring strips. Framing for the ends of the wall is pieced in, with a sloped section of top plate nailed in place, along with end studs and any other studs that fall on the 16-in. o.c. layout.

The stair-chase wall needed a sloped portion at one end as well, but there's an easy way to handle framing

The top plate of the bathroom wall is shorter to fit between the angled ceiling sections. Plumb each end before nailing the top plate. Then, after the wall is secured, fill in the ends with a slanted plate section and studs with bevel cuts at the top.

Measure between the posts to get the actual length of the bathroom wall (right). Cut the bottom plate and mark the layout (below).

Fill in studs over to the angled part of the ceiling. Then cut a top plate for the angled section and measure over for the stud layout. Measure the length of the fill-in studs in place.

Inspect all the interior framing to make sure there is adequate nailing for the drywall. Here, a nailer is added where the cheek wall meets the dormer gable.

To locate a wall in spaces such as the stair-chase wall, install the bottom plate and then plumb up to the ceiling to locate the top plate. Install the top plate and measure between the plates to determine the length of the studs.

in unusual situations such as this. Nail the bottom plate to the deck, then put a level on a straight 2×4 and plumb up to locate the position of the top plate. Cut a length of 2×4 for the top plate, and nail it securely to the plumb mark, but leave it long. Now plumb up from the end of the bottom plate to mark the corresponding end point on the top plate, and cut the top plate to that length. Fill in the studs on the layout. At the sloped section of the ceiling, nail in a short section of plate. Measure over for the stud layout on the sloped plate, and then measure up from the bottom plate for the length of the stud. Cut and install that stud as well as the end stud to complete the wall.

Mop up odds and ends

After all the interior partitions are complete, make sure that there is a drywall nailer at every edge of every wall and ceiling. No matter how careful you've been at earlier stages, I guarantee that there will be places where you need to add nailers. The second floor and its angled ceilings demand especially close attention.

Framing Kneewalls

Kneewalls close off the less usable portion of an area where the rafters meet a floor. They are simply short walls that have an angled top plate nailed to the underside of the rafters. The 2×4 kneewalls in this house run from the gable walls to the dormer cheek walls.

The plans specified that the walls be placed 5 ft. in from the rafter plates. So mark that distance in from either end and snap a line between the marks. Cut and install the bottom plate, then build partition backers at both ends so you have something to anchor the knee-walls to. (By the way, the nailer for a backer doesn't have to be a continuous length of 2×. This is a great place to use up short lengths that would otherwise end up in the scrap pile.) Now plumb up from the plate to the rafters at either end, and mark those points. Then snap a guide line for the top plate.

The top plate is a little different because it has to be nailed to the underside of the rafters. That means it has to be made from stock wider than 2×4s and the top edge of the plate has to be given the plumb angle to match the 10:12 pitch roof. So rip the angle on the edge of a 2×6, and nail it to the guide line on the rafters. Studs line up directly below the rafters, so the layout is easy, though each stud has to have an angled top so it will fit against the plate. In the middle of each kneewall, frame a small access panel using a nonstructural header for the open-ing. The plans set the width of the panels, but the height was determined on site.

To locate the kneewalls, install the bottom plate and plumb up to locate the top plate (above). Snap a guideline for the top plate, rip a plumb cut on the edge of a 2×6, and nail the plate to the line (left).

The kneewall studs must be cut at an angle to fit against the plate. An access panel in the middle of the wall should be built just like any door opening in a nonstructural wall.

One of the places we had to attend to was where the angled ceiling extended beyond the stair-chase wall. Not only did we have to piece in a sloped nailer at the posts but we also had to install a horizontal nailer along the top of the little triangle. Other places to watch out for are where the ceiling meets the gable walls. Drywall nailers must also be added in the stud bay below each gable rafter.

Finally, go back to all the doorways on both floors and cut out the plates. Otherwise, if you're like me, you'll soon trip over one and fall flat on your face. After removing the plates, drive a couple of 16d nails through the end of the plate and into the deck to anchor the sides of the door openings in place.

At the posts that support the dormer headers, the angled ceiling extends past the stair chase, so we added short 2× nailers for the drywall.

To remove trip hazards, cut out the plates at each doorway with a handsaw or reciprocating saw (left). Toenail the cut ends to the deck to anchor them (below).

Exterior Trim and the Front Porch

On one of the very first houses I framed, I recall arriving at the site on a beautiful spring morning the day after we'd finished sheathing the roof. The head of the crew pulled in behind me, handed me a coffee, and said, "Hey, this place is beginning to look like a house!" Indeed, it was the first time the house had the shape it would have for the foreseeable future.

Our project house is at that stage as well, and that means the next task is to install the exterior trim and build the front porch. Once these are done, the installation of roofing, siding, doors, and windows will make the house shell weathertight.

Ordinarily the porch framing should be done before the trim. Unfortunately, bad weather delayed digging and pouring the porch piers so we skipped ahead to the trim. This approach took a bit of planning because we had to figure out where the porch framing would be to determine where the trim should end. It wasn't an ideal situation, but serves as a reminder that you just have to work around whatever Mother Nature throws your way. But wait for good weather whenever you can—building a house for yourself doesn't expose you to the hard deadlines of a professional crew anxious to tackle the next project.

Installing Exterior Trim

At this point there's trim on the gable ends but all the corners of the house are still raw plywood edges, and the rafter tails are still exposed. But if you've ever sketched an object, you know that emphasizing its outline makes it seem to pop off the page. Exterior trim does a similar thing for a house.

Exterior trim on some houses can get really complex and is often beyond the realm of mere mortal framers. In these cases, finish carpenters complete the exterior details. With this project, however, the exterior finish details were fairly simple and did not require such assistance. Another thing that simplified the work was using preprimed trim (trim material coated on both sides and both edges with factory-applied primer). Although it costs a little more, preprimed trim provides added protection from the elements and improves the stability of the wood. It also gives the paint contractors a head start on their job, and keeping the subs happy is very important. But before you tackle the trim, put up wall jack scaffolding to make the process a lot safer and easier (see "Installing Wall Jacks" on p. 205).

Trim emphasizes the shape of a house, and should be installed before the siding and roofing.

STEP-BY-STEP

Installing Exterior Trim

1 Install the soffits and continuous vents.
2 Install the fascia.
3 Install outside corner boards.
4 Install the frieze.
5 Install the inside corner boards.
6 Install the returns.
7 Install the dormer trim.

Eaves Trim Details

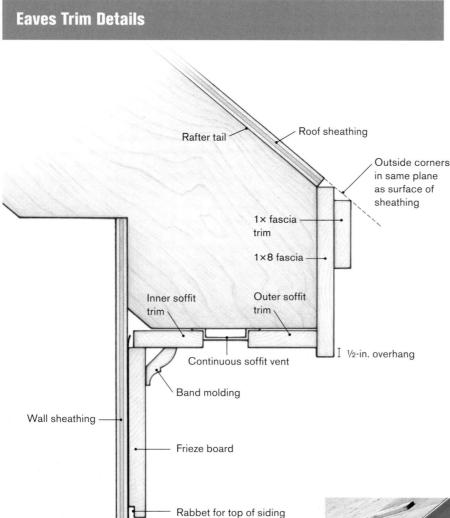

Roof sheathing

Rafter tail

Outside corners
in same plane
as surface of
sheathing

1× fascia
trim

1×8 fascia

Inner soffit
trim

Outer soffit
trim

Continuous soffit vent

½-in. overhang

Band molding

Wall sheathing

Frieze board

Rabbet for top of siding

Felt paper or waterproof
membrane

Install the soffit

The eaves trim begins with installation of the soffit (the horizontal assembly under the rafter tails). I've built soffits a dozen different ways over the years, but the method I'll describe here is simple, and it provides for ample ventilation, which is essential for maintaining the health of your house.

The first piece to go on is the outer soffit trim. We ripped a 1×6 in half for this piece, so it turned out slightly wider than a 1×3. Before measuring the length of the soffit, temporarily slip a scrap of 1× finish lumber between the sheathing and the bottom edge of the rake trim. The unsupported lower edge of the rake trim (remember, it's a 1×8 on top of a 1×3 furring strip) probably bends slightly toward the house. The scrap eliminates the bend and plumbs the rake trim so you can get an accurate measurement for the soffit.

After cutting the outer soffit trim to length, align it with the plumb cuts on the rafter tails and tack it in place every three or four rafters. Sight down the length to make sure it's perfectly straight. On the sticking-out-like-a-sore-thumb scale, wavy or bowed eaves trim

Nail the outer soffit trim to the bottom edge of the rafters, but nail just the outer edge for now. (If local codes require hurricane ties on the rafters, as here, install them before building the soffit.)

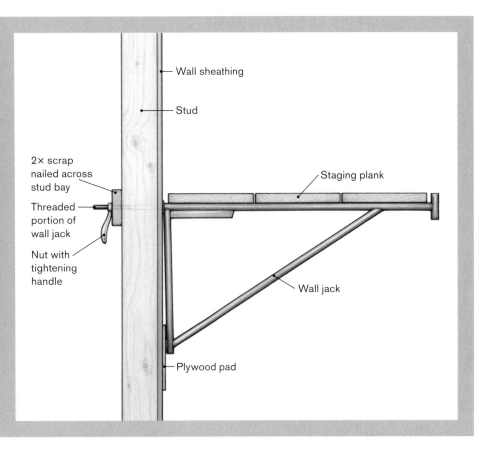

SAFETY

Installing Wall Jacks

The best way to finish the eaves of a house is to work from scaffolding—working directly from a ladder is far less convenient and not as safe. The easiest and fastest way to provide scaffolding is with wall jacks. The horizontal portion ends in a threaded rod that extends through a hole drilled in the wall sheathing in between two studs. The bolt then goes through a scrap of 2× that spans the stud bay from the inside, and a nut with a handle secures the jack in place. On the outside, it's a good idea to slip a scrap of plywood under the bottom flange to distribute the pressure on that flange more widely.

Labels in diagram: Wall sheathing; Stud; 2× scrap nailed across stud bay; Threaded portion of wall jack; Nut with tightening handle; Staging plank; Wall jack; Plywood pad

---

ranks right up there with crooked rakes. When you're satisfied, nail off just the outside edge of the piece using light-gauge stainless-steel nails with heads.

Install the continuous soffit vent next. The vent we used is a vinyl strip that can be cut to length easily with a utility knife. Slip one leg of the strip between the outer soffit trim and the bottom of the rafters, then nail off the inner edge of the trim to hold the vent strip in place. Cut the strip into lengths so that butt joints meet at the center of a rafter. Next, install the inner soffit trim over the other leg of the vent, using two nails at every rafter. Note that there's a healthy gap between the inner soffit trim and the wall sheathing (see "Eaves Trim Details" on the facing page). The gap gives you plenty of room to fit the soffit pieces into place and will be covered later by the frieze board.

Attach the fascia

The fascia that caps the ends of the rafter tails has two parts: a 1×8 fascia and a 1× fascia trim made from a 1×6 ripped in half lengthwise. Cut the 1×8 to length and have someone hold one end as you nail it in. The

Cut vinyl vent material with a utility knife, and then tuck one side under the outer soffit trim. Nailing the inside edge of the trim secures the strip in place.

Attaching Trim

Unlike framing lumber, trim stock is exposed to the weather so installation calls for a somewhat different set of techniques. Similarly, the fasteners you use have to be weather resistant as well as unobtrusive. Here are some general tips for working with 1× trim stock.

- **Joints should fall over framing.** Whenever possible, join lengths of trim so that the ends land over solid wood, then drive the nails into the framing through both abutting boards to hold the joint tight.

- **Caulk butt joints.** Simple butt joints can be used for horizontal trim such as eaves and soffits because the joints will be somewhat protected from the weather and gravity won't draw moisture into the joint. Caulk the joint with a paintable, exterior-grade caulk.

- **Protect exposed joints.** Exposed joints such as those on rake trim and corner boards should be joined with scarf joints so that water will shed naturally to the outside of the trim. Mate the boards so that the higher one overlaps the lower one.

- **Use the right nails.** All exterior trim should be fastened with stainless-steel nails. Stainless steel won't rust and discolor the way ordinary steel can. We used medium-gauge ring-shank nails with small flat heads. The ring shanks give the nails superior withdrawal resistance, and headed nails hold better than finish or casing nails though they're somewhat more visible in place.

- **Set the nails flush.** If you're using a nailer, set the compressor so that the nailer drives the nail heads flush with the surface of the wood. House paint seems to last longer that way; filled nail holes don't hold paint as well as exposed nail heads.

fascia overhangs the soffit by ½ in., so check this distance as you nail. The overhang hides any irregularities in the plane of the soffit and creates a natural drip edge to keep water from running back onto the soffit. Nail the fascia to the edge of the outer soffit trim as well as to the ends of the rafters.

The fascia conceals the ends of the rafter tails and closes up the soffit area (above). Measure from the bottom edge of the fascia to the soffit to keep the overhang consistent (right).

Simple Rake Detail

The overhang rake detail chosen for this house is common in this part of the country. The extended rake hides the end of the gutter and the butt joint between the rake and facia is less emphasized.

It's worth noting that the installation of the fascia will point out how thoroughly you thought out the rafter tails when you framed the roof. In this case, we cut the tails so that a full 1×8 fascia could be installed without having to be ripped to width and so that its top edge would step down neatly in line with the surface of the roof sheathing. The fascia trim (which we added later) has to step down even farther to stay in plane with the roof sheathing (see "Eaves Trim Details" on p. 204). If the fascia and trim are installed too high or too low, there will be an unsightly flare along the lower edge of the roof shingles. By the way, we held off on installing the fascia trim only because the stock wasn't on site when the fascia was installed. It can be installed at any point after the fascia is up.

Once the fascia is in place, you can turn to a bit of unfinished business and cut the rake trim to length. The plans for this house called for the rake trim to extend past the fascia by 6 in., a common trim detail in this area. In other parts of the country you might find different details instead (see "Simple Rake Detail" above). The extended rakes will be covered by a narrow strip of roof shingles later. First mark a plumb line on the rake 6 in. out from the fascia. Then cut along the line carefully with a circular saw. Whether you cut from the top or the bottom of the board

depends on being able to see the line while holding the saw and standing safely on the staging. At this point in the framing process, you may feel comfortable enough to make this challenging cut with a circular saw. But if the task seems scary or uncomfortable to you, use a handsaw instead.

Assemble and install the corner boards

Corner boards for outside corners can be a little tricky. Both sides of the corner appear to be the same width, but the corners are actually formed by two boards of different widths. Building them is easy enough—nail two 1× boards together at the edges. But in order for them to measure the same dimension on both sides of the corner, you first have to rip ¾ in. off the edge of one of the boards. Save the ripped piece to use later—it makes a perfect inside corner trim.

To cut the rake trim to length, measure over from the fascia and draw a plumb line. Then cut to the line.

Measure and cut both sides of the corner, then nail them together along one edge. Note that one side is longer and has a pitch cut that will fit under the rake trim.

For most outside corners, one side extends up under the rake trim, so make a pitch cut on one end of that board. (Use the narrower piece for this side so that the wider piece extends to the end of the soffit above.) The board with the pitch cut has to be several inches longer than the other side of the corner. When you measure the lengths for the two sides, make them long enough to extend a few inches past the bottom edge of the wall sheathing. They will be cut to exact length in place, after the siding is installed. Assemble the corner by nailing the edges together with 2-in. ring-shank stainless-steel nails every foot or so.

Before installing the corner trim, wrap the corner of the house with a strip of building felt wide enough to extend past the edges of the trim by several inches. You can staple the felt paper directly to the sheathing,

Felt paper helps weatherproof the corner of the house underneath the corner boards, and it ties into the housewrap on the walls, but instead of stapling the paper to the house, you can attach it to the back side of the corner board.

but keep the paper as flat as possible—bulges will keep the corner from fitting tight. A different approach is to apply the felt paper to the back side of the corner assembly before it goes on, as shown in the photo above. Fold the felt paper in half, then use a wood scrap to push the fold into the corner of the assembly as you drive staples to hold it in place. Attaching the felt paper to the corner assembly makes sense if you're working in really windy conditionsand makes the entire corner, paper and all, go in with one step.

When ready, install the corner by sliding the long side under the rake. Have one person push the corner assembly against the house while you nail it in on both faces every 12 in. to 16 in.

Install the frieze board and band molding

The next finish detail is the 1×8 frieze board—that's the horizontal band below the soffit. The frieze closes the joint where the soffit meets the sheathing and caps the top edge of the siding. Cut the frieze board to length and cut a rabbet ½ in. wide and ¼ in. deep in the back edge of the board to capture the top edge of the siding. A router with a bearing-guided bit makes quick work of this step, but you can cut the rabbet on a tablesaw instead.

To install the frieze, start by tacking a strip of felt paper directly below the soffit. Make sure the paper is wide enough so that you can tuck the housewrap

Have one crew member push the corner board assembly against the house while you nail it in from either side as you work your way down from the top.

Cut a rabbet on the lower edge of the frieze board so it will fit over the siding. A bearing-guided router bit makes the best rabbet.

under it in preparation for siding. Snug the frieze up against the soffit, and nail it to the wall with three medium-gauge stainless-steel nails every 16 in.

Inside corners can go in after the frieze is installed. The inside corner gives the siding a flat termination point instead of having to butt into the siding on the adjacent wall. Again, fold a strip of felt paper in half, stick the fold into the corner, and staple both sides to the sheathing. Remember that ¾-in. piece you saved when ripping the corner boards? Cut it to length and nail it in over the felt paper in the corner. Nail the strip from both sides with the same stainless-steel nails used for the rest of the trim, so that it stays tight in the corner.

Install the returns

The trim piece that finishes the end of the soffit is called a return, and the returns on this house are

After stapling a strip of felt paper into the corner, nail in the frieze board. The upper-most piece of felt should always overlap the piece below to shed moisture.

An inside corner serves as a termination point for siding.

Returns cover the exposed edges of the soffit on the gable end. Measure from the rake along the edge of the corner board and add ½ in. to provide an overhang (top right). Then cut and fit the return (bottom right).

simple and unadorned. Because of their triangular shape they are often called *pork chop returns*. To get the vertical length of the return, measure down to the bottom edge of the soffit and add ½ in. to provide an overhang that matches that of the fascia. Then measure the horizontal distance back to the rake. The diagonal on the return should match the roof pitch.

Nail the pork chop to the corner board as well as to the ends of the soffit trim, making sure to keep the overhang consistent with the fascia. Drive a single nail through the point of the triangle up into the rake trim. Just be sure to drive the nail straight so that it doesn't shoot out the face of the trim.

A decorative band molding finishes off the corner between the soffit and the frieze and visually softens the intersection of the two surfaces. Drive 2-in. light-gauge finish nails every 12 in. to 16 in., alternating the direction of the nails so they angle into the soffit and into the frieze as well. Be sure to install the molding right side up: The convex part goes on top to make the installation architecturally correct. The final task is to install the fascia trim. Use the edge of a triangle square to align the outside corner of the trim with the roof plane before you nail it in, as shown in the bottom photo at right. This tactic will prevent the trim from interfering with the roofing.

Install the dormer trim

The trim for the dormers is pretty much the same as elsewhere except for a few details where the dormer meets the main roof. Apply self-adhesive waterproofing membrane to the main roof sheathing where the eaves of the dormer end (see the top photo on p. 212). Some framers leave this detail to the roofers, but it's much easier to put the membrane in before the trim goes on. This membrane is much thicker and tougher than felt paper and has an adhesive backing that makes it stay in place without the need for nails. A word of caution, though. The adhesive is not effective in cold weather, and staples may be needed to hold the membrane in place. The membrane is self-healing so that it seals around any fasteners driven through it.

The fascia boards on the dormers die into the main roof with a pitch cut. Lay a length of 1× on the roof as

A strip of decorative band molding softens the corner between the soffit and the frieze board.

A secondary frieze board is installed in line with the roof plane. A triangular square guides the alignment.

STEP BY STEP

Framing the Porch

1 Build the porch support beam.
2 Install the porch floor framing.
3 Install LVL support beams.
4 Install the porch ceiling.
5 Install the rafters.
6 Sheathe the porch roof.

Before you trim out the dormers, add waterproof membrane where the roofs intersect to provide leakage protection (above). Then use a 1× scrap to space the end of the eaves trim off of the roof to allow for the roof shingles (right).

personal preference is to use waterproof membrane to protect these areas, but the crew on this house opted for the time-tested and reliable defense of lead flashing. Lead is soft and flexible enough to be molded and folded to shed water around the entire corner, but it's also a toxic material and should be handled with extreme care. Use a piece of scrap wood and a hammer to press the flashing into place. It has to sit on top of the roof shingles so hold it off the roof plane slightly. The felt paper for the dormer corners should extend over the top of the lead flashing, and the corner boards themselves should be spaced off of the roof sheathing like the eaves trim above.

Framing the Porch

To me, no feature adds a more friendly and inviting note to a house than a front porch. In hot weather, a porch creates a cool shady place to sit, and, like the one on our project house, a porch shelters the front door from all but the worst wind-driven weather.

The front porch is the last framing project for this house. Its details and dimensions (22 ft. long, 4 ft. 3 in. wide) come from the first-floor house plans and the section drawings (see "Porch Framing Details" on the facing page). Note that the 6:12 pitch of the porch roof is shallower than that of the main roof. This provides a bit of extra headroom under the porch rafters.

a spacer to hold the eaves trim off the roof sheathing. This space will allow the roof shingles to slide easily underneath the trim, plus a little extra so the trim isn't in direct contact with the roofing.

Another area of concern occurs where the front corners of the dormer meet the main roof. These corners are notorious locations for water leakage. My

Porch Framing Details

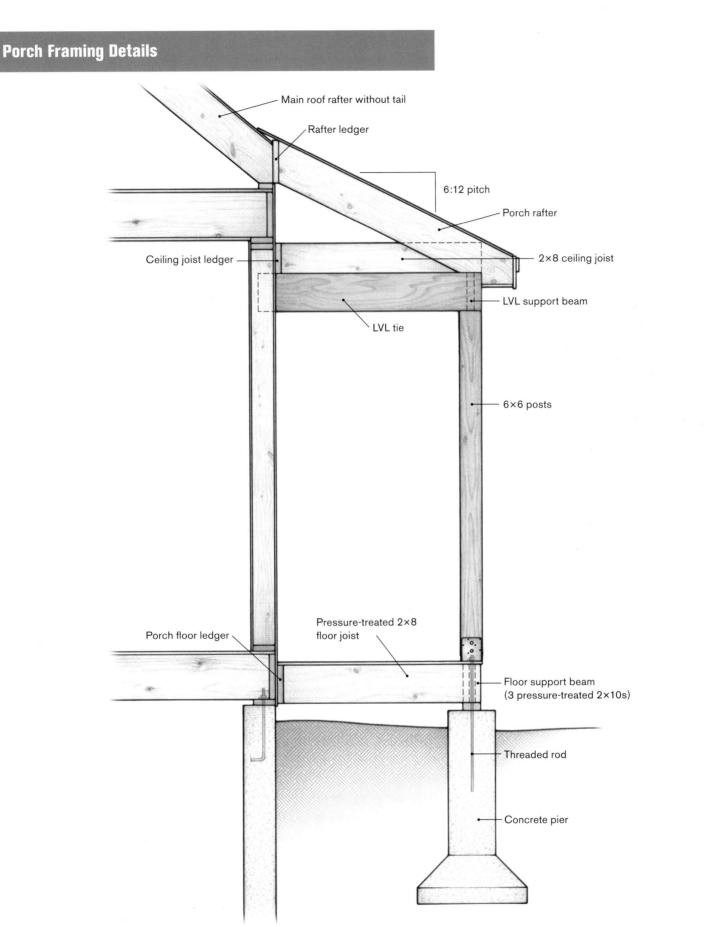

Main roof rafter without tail

Rafter ledger

6:12 pitch

Porch rafter

Ceiling joist ledger

2×8 ceiling joist

LVL support beam

LVL tie

6×6 posts

Porch floor ledger

Pressure-treated 2×8 floor joist

Floor support beam (3 pressure-treated 2×10s)

Threaded rod

Concrete pier

Build the porch floor support beam

The outermost side of the porch sits on a built-up lumber beam resting on four concrete piers. Some contractors dig and pour porch piers when the main foundation is installed, but four pillars of concrete sticking out of the ground during framing invite accidents and could be damaged as construction goes on around them. Instead, we installed the piers a few days before porch construction started. That's all the curing time necessary for the concrete to be strong enough to work on.

The layout for the porch is taken directly from the position of the piers and then transferred back to the house, not vice versa. This approach may seem counterintuitive, but the piers are, well, cast in concrete so they are where they are. Where exactly the porch framing meets the house is less important than centering the porch posts directly over the piers.

Porch framing begins with the support beam, which is simply three pressure-treated 2×10s nailed together with five 16d nails every 16 in. (see "Porch Floor Beam Details" below). Gaps in the middle layer make room for threaded rod that secures the beam to the piers. To assemble the beam, place the first layer on top of the piers, and lay out the overall length of the porch as well as the post locations before cutting the stock to length. This information can be found on the plans. You'll need two boards for this distance, so make the joint between them land over a pier. Mark the post centerlines on the edge of each pier as well. Then place a 4-ft. drywall square against the sheathing of the house and transfer the end post positions back to the house wall. Cut the board to length while it's sitting on the piers.

Porch Floor Beam Details

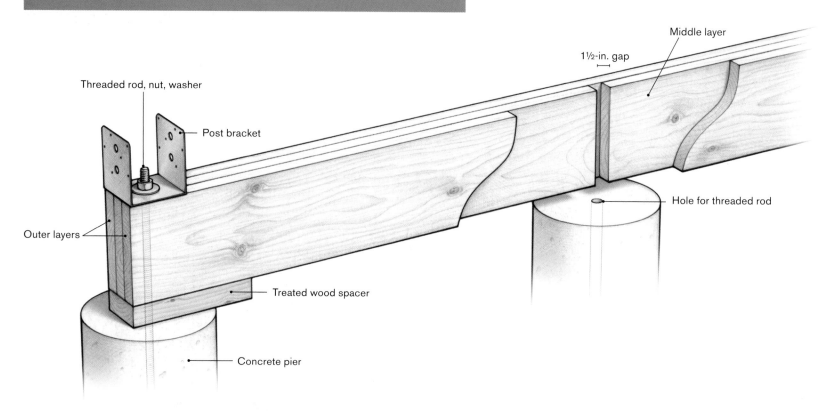

- Threaded rod, nut, washer
- Post bracket
- Outer layers
- Treated wood spacer
- Concrete pier
- Middle layer
- 1½-in. gap
- Hole for threaded rod

Set the first layer of the support beam on the porch piers, and mark the length of the porch as well as the post locations (above). Transfer the marks to the piers and use a large drywall square to mark the locations on the house (right).

Plumb up from the post locations marked earlier and measure across for the length of the porch.

For the second (middle) layer, space the boards to leave 1½-in.-wide gaps over the center of each post layout—these spaces will accommodate threaded rod to connect the post brackets to the piers. On the third layer, bridge the gaps and stagger the seams between boards so they are not in line with the seams on the first layer.

Then on the wall sheathing of the house, draw a plumb line up from one end of the porch and mark the overall width of the porch just below the rafters. Mark the key measurements now because a strip of water-proof membrane will soon cover layout marks at the base of the wall.

Set the brackets Set the beam to one side and mark the distance from the house to the center of the post on the piers at each end of the porch. The post brackets will be able to compensate for slight variations, but try to get the measurements as close as possible. Snap a chalkline between the marks across the tops of the other piers. With the side-to-side post layout marks from before, you now have cross-hairs for drilling the bolt holes.

Mark the width of the porch to the center of the posts on top of the end piers, and then snap a line between the two marks. This line crosses the post lines made before to pinpoint the bolt locations.

Attaching Brackets to Concrete Piers

On this porch, threaded rod connects each pier to a metal post bracket that sits on top of the support beam. To install the rod, drill a hole into the top of each pier about ⅛ in. larger in diameter than the threaded rod. Use a masonry bit chucked into a hammer drill, which spins the drill bit as it also drives it into the material, making quick work of drilling holes in concrete.

When the hole is drilled as far down as the bit will go, clean it out thoroughly with compressed air (be sure to protect your face from flying grit), and squirt in industrial-strength epoxy glue specially made for this purpose. Then push the rod into the hole, and tap it carefully to the bottom to avoid damaging the threads. Most epoxies set up fairly quickly, although some may take longer in cold weather. Check the epoxy before setting the beam. I like to make sure that it has at least skinned over and started to set before fitting the beam.

To anchor the bolts in the piers, first drill holes with a masonry bit chucked into a hammer drill. After filling the holes with industrial epoxy (above left), carefully tap threaded rod all the way to the bottom of the hole (above right).

Carefully slide the porch support beam over the threaded rod. Note that pads have to be measured and installed to raise and level the beam at the correct height for the porch floor.

Secure the beam With a third person to help hold the middle of the beam and guide it over the bolts, carefully slip the beam into place. The top of the porch support beam is also the height of the porch floor minus the thickness of the flooring material, and, at this point, the height of the porch beam still has not been set. To set the height and to level the beam, cut and place treated-wood spacers at every pier. You might wonder why the piers weren't just installed to the right height in the first place, but believe me, that's hard to do and ends up being very time-consuming. Besides, piers can settle and move slightly after they're in place. So the idea is simply to get the piers in the ballpark for height and then use permanent spacers to lift the beam to the correct height.

To determine the thickness of the spacers, level over from the house at one post position. Then level the beam from that post to the others and measure for the spacers. To measure for the spacers, pry or lift the beam to a level position. Shims and 2× blocks can hold the beam temporarily until you make the spacers. To

make the spacers, rip lengths of pressure-treated 4×4 stock to the right thickness, then drill holes in the spacers to accommodate the bolts. Lift the beam off the piers, and slip the spacers over the bolts before dropping the beam back down. Now slide the metal post brackets over the bolts and hand-tighten the nuts to hold the brackets in place. When the brackets are set with the nuts on the bolts, cut off the excess threaded rod with a hacksaw, a grinder, or a reciprocating saw fitted with a metal-cutting blade.

Attach the porch floor ledger

Now turn your attention to the porch floor ledger, a pressure-treated board connected to the house to support the other end of the porch joists. This is a critical connection in terms of safety so be meticulous with the details (see "The Ledger Connection" on the facing page).

To protect the framing and sheathing of the house against water damage, install a layer of waterproofing membrane over the bottom portion of the walls before

Before installing the porch floor ledger, attach a layer of self-adhesive waterproof membrane to the wall sheathing.

After the ledger is tacked in place, drive ledger fasteners through the ledger and into the house framing with an impact driver.

The Ledger Connection

Ledgers for porch floors and decks have been a major weak point in house construction over the last few decades because of poor connections to the house. If this connection fails, people can get injured. So never fasten a ledger just to the sheathing or just with nails; always screw it or bolt it to the floor framing of the house. By the way, building codes have recently gotten extra fussy about deck and porch ledger connections. Some specify that ledgers be through-bolted to the floor framing and may even require the installation of specialized metal ties to connect the floor joists of a porch or deck directly to the floor joists of the house. The lesson? Always be sure the installation of a deck or porch ledger meets or exceeds your local code requirements.

installing the ledger. The membrane should extend several inches above the ledger, so snap a line there to ensure proper alignment.

Once the membrane is in place, level over from the beam at both ends and snap a chalkline for the top of the ledger. At 22 ft. long, the ledger has to be installed in pieces. Set the first length to the chalkline and tack it in place with a few 16d nails. Measure the balance of the ledger, cut that board, and tack it to the line as well.

Now go back and drive two ledger fasteners every 16 in. along the length of the ledger. Ledger fasteners are a type of lag screw made especially for this purpose. They're long enough (3⅝ in.) to go through the ledger, the wall sheathing, and into the rim joist.

Install the first joists

The pressure-treated floor joists set the width of the porch. First, mark an identical joist layout on both the ledger and the beam. To measure for the joists,

Install floor joists at the ends and in the middle to hold the support beam straight and in position.

Before building the porch ceiling and roof, make temporary scaffolding brackets out of 2×6s anchored to the house. The brackets support a heavy-duty staging plank.

fine-tune the beam to the proper distance from the house and measure between the ledger and the beam. (There's enough play around the securing bolts to move the beam in or out slightly as needed.) Then cut three joists and install them at the two ends and in the middle to lock in the beam at the width of the porch. Secure the joists with toenails for now (joist hangers will be installed later on). Having only three joists at this point also makes it a lot easier to walk around while the rest of the porch is built.

By the way, while the plans for this house called for 2×10s for the beam and for the ledger, 2×8s spaced 16 in. o.c. are big enough for the joists. And in case you're wondering, the porch decking was specified as ¾-in.-thick Ipe deck boards. Contractors' boots can ruin deck boards very quickly, however, so the decking was installed at the same time as the floors inside the house to minimize damage. After the porch joists were

installed, we covered them temporarily with plywood for a flooring surface.

Build the roof

The roof of the porch consists of rafters and ceiling joists supported by the house and by an LVL support beam. Unlike the house rafters, however, the rafters for the porch rest against a ledger at their upper ends, instead of against a ridge beam. This configuration is known as a shed roof—a roof pitched in one direction only.

Scaffolding makes it a lot easier to install the porch roof framing, so we built simple brackets out of 2×6s and braced them back to the house with a diagonal 2×4. Because these brackets were being used for only a very short time, and because the ground was frozen solid, we just set the end of the 2×6 on the ground. Ordinarily, it's best to set the bracket post on a 2× pad to keep it from sinking into the ground. We set our heavy-duty staging plank on top of the brackets.

Install the roof support frame The porch roof frame is supported by four posts, each one placed directly above a pier. But rather than install the posts

Porch Tie Details

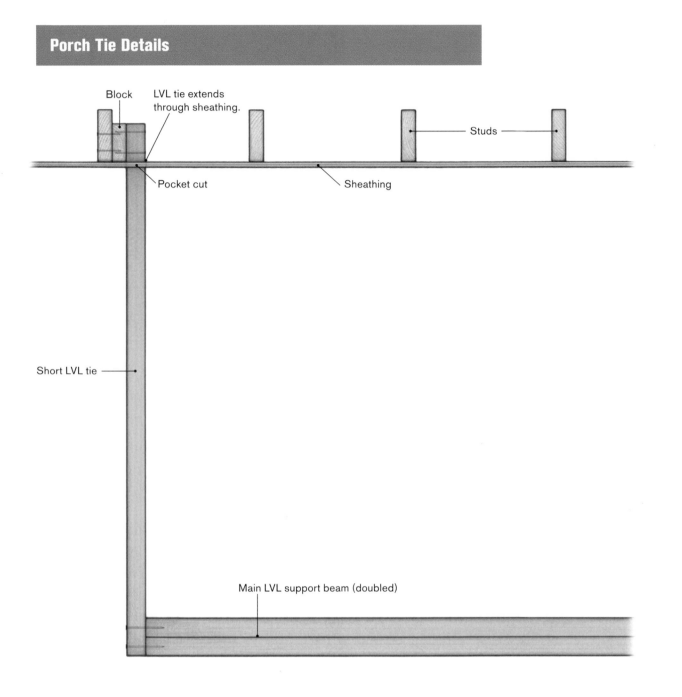

Block

LVL tie extends through sheathing.

Studs

Pocket cut

Sheathing

Short LVL tie

Main LVL support beam (doubled)

first and build the roof structure on top of them, it's actually easier to build a portion of the roof system first and then slip the posts into place beneath it. The key element of this porch roof is an LVL frame: a long doubled-LVL beam that supports the lower end of the rafters and single short LVL tie at each end of the beam that connects the beam to the house and prevents it from twisting. You'll find it much easier to install the LVL ties before the beam.

There are various ways we could have connected the LVL ties to the house but we opted to connect each

one to the side of a wall stud (see "Porch Tie Details" above). That meant we had to plunge-cut a pair of rectangular holes in the sheathing just big enough to let the LVL tie slip through.

The outer edges of each hole are in line with the ends of the porch, which you marked earlier on the sheathing. The height of the hole can be determined from the plans because it's the same height as the ceiling joist ledger, so locate that first. Measure up from the porch floor framing, then fine-tune the measurement so that a chalkline snapped across the

ANOTHER WAY TO DO IT

Cutting Holes in the Wall

If the idea of plunge-cutting wall sheathing with a circular saw intimidates you, drill holes in the corners of the layout and cut between the holes with a reciprocating saw. The results will be the same. There's usually more than one way to do anything when building a house, so find the method you feel most comfortable with and that will likely be the safest one for you.

Cut a hole in the sheathing so that each LVL tie can be slipped through and fastened to the side of a stud. The other end of the LVL rests atop a temporary post.

Measure up from window headers at both ends of the porch, then snap a line between the points. The line represents the bottom edge of the porch ceiling joist ledger and keeps the porch ceiling parallel to the tops of the windows.

wall will be exactly parallel to the tops of the windows—if it's off even a little, the difference will be obvious because the porch ceiling will be close to the top of the window trim.

Once you know the location of the holes for the LVL ties, lay them out and carefully plunge-cut through your layout (see chapter 4). Then slide a short LVL into the hole and tack it to the sheathing with a toenail from the outside. After the LVL is in place, add blocking if necessary between the stud and the LVL inside the wall to make the connection permanent.

Support the free ends of the ties with temporary posts made out of doubled 2×6s (they have to be strong enough to support the weight of the main LVL beam until the permanent posts can be set). Nail on short diagonal braces to keep the ties square to the house wall and parallel to each other.

The long beam is two 10-in. LVLs joined together by pairs of 3⅝-in. lag screws spaced every 16 in. (Drive them through the surface that will face the house so that the heads won't interfere with the exterior trim later.) Lift the beam into position, slip it

The long LVL beam fits between the shorter ones and rests on a temporary post in the middle. Tack the beam to the post to keep it from being knocked out.

The ledger for the porch ceiling connects to the house framing above the line snapped for the beam (top). Install a few ceiling joists to hold the LVL beam straight (above).

into place *between* the LVL ties, and secure each end with five 16d nails. Then add another temporary post in the middle of the beam.

Install the ceiling joist ledger Align the 2×8 ceiling joist ledger to the chalked line you made earlier, and tack it in place with a couple of 16d nails every 4 ft. or so. Like the floor joist ledger below, the ceiling joist ledger must be screwed to the house framing with 3⅝-in. ledger fasteners. But because it will be protected by the porch roof, the ceiling joist ledger doesn't have to be treated lumber, and you don't have to place waterproof membrane behind it. Mark the 16-in. o.c.

layout for the ceiling joists on both the ledger and on top of the LVL beam. Then install a joist every 5 ft. or so to hold the beam straight. Hold off on installing the remainder of the joists so that they won't be in the way as the roof framing is laid out.

Calculate the rafters To figure out the cuts for the porch rafters, use the same basic process you used to calculate rafters for the main house and dormers. But because the porch roof is only a single plane (a shed roof), the run of the rafters is the distance from the face of the wall sheathing to the outside edge of the LVL beam: in this case, 4 ft. 3 in. The length per foot of

Cut and test-fit a porch rafter. Mark the peak at both ends of the dormer wall. Note that the peak and the rafter tail still need to be laid out and cut before the rafter can be used as a pattern.

Cut back the main roof sheathing to expose the ends of the rafters over the porch.

Snap a line between the marks made on the sheathing for the rafter ledger.

run for a 6:12 pitch is 13.42 in., which gives a theoretical length for the porch rafters of 57.035 in., or about 57¹⁄₃₂ in. In framing lingo, we'd just call that "57 inches strong" and cut the board a hair longer than 57 in. Before accounting for the ledger or cutting the overhang, make a test rafter with just the plumb cut and the bird's mouth and try the fit.

To make the pattern rafter, start by adjusting the top plumb cut of your test rafter. The length of the test rafter was calculated from the sheathing. However, the ends of the porch roof rafters will actually rest against a 2× ledger, not against the sheathing. That just means you have to make the end plumb cut 1½ in. shorter on your final rafter layout to account for the thickness of the ledger. You did the same thing when you accounted for the thickness of the ridge board in laying out the house rafters. After you adjust the plumb cut, lay out the rafter tail. It has the same horizontal length as those on the rest of the house.

The porch roof stops in line with the outside wall of the dormer at the right side of the porch. But at the left side it blends into the main roof. For various reasons, the tops of the porch rafters on that side can't be nailed directly to the roof sheathing or to the ends of the shortened house rafters. Instead, they must be attached to a ledger. First, cut away the roof sheathing at the ends of the shortened rafters. (If you nailed off the bottom edge of the sheathing earlier, remove those nails with a cat's paw before you start sawing.) It's okay to rip a little extra to make sure the sheathing doesn't interfere with the ledger. To set the height of the ledger, mark the end of the test rafter at either end of the dormer and snap a chalkline.

Toenail the rest of the ceiling joists to the ledger. Then set plywood sheathing on top of the joists as a temporary platform for building the roof.

With all the figuring done for the porch rafters, install the rest of the porch ceiling joists. For now you can just toenail them into the ledger and into the top of the doubled LVL beam.

Attach the rafter ledger The porch roof was too long to make the ledger out of a single board, so we split it into two sections, one that we attached to the dormer wall and the other that we attached to the plumb cuts of the shortened rafters. Tack the boards in place and then fasten them to the house framing with 3⅝-in. ledger fasteners. To hold the rafter tail ledger at the right height, set a scrap of 2× on top of the ceiling joist ledger to hold up one end. Then align the other end with the ledger on the dormer, tack it in place, and screw that section to the ends of the rafters with the same ledger fasteners as before.

The porch rafter ledger is in two sections. The first lags into the dormer framing (top). A block on top of the ceiling ledger holds the second section at the right height while it is attached to the ends of the rafters (above).

Install the porch posts

Before the roof can be framed, we need to install the permanent posts. These posts could have gone in earlier, but the 6×6 stock didn't arrive until the middle of the process. Rather than wait, temporary posts served the crew just fine until that stock was delivered.

Measure from the post bracket to the bottom of the LVL beam at each post location. Then cut the 6×6 posts to length (see "Cutting Large Square Stock" on p. 226). Before putting in the posts, cut a 2×6 prop about 1 in. longer than the post height. Starting at one end, tap the prop in next to the temporary post until it raises the LVL beam slightly. Then pull out the temporary post.

While the prop is still holding the LVL slightly high, set the post on its bracket and slip it into place

Cutting Large Square Stock

The posts for the front porch of this house are made from 6×6s, which have an actual dimension of 5½ in. on each face. That's too deep for the blade of a 7-in. circular saw (the maximum cut depth is typically 2 ¾ in.). So here's how to cut the posts.

It's crucial to have the blade perfectly square to the saw base, so check that first. Then square a line across one face of the post and extend it down the two adjacent faces. Roll the post so that one adjacent side is facing up and cut to that line with the sawblade set at maximum depth. Now rotate the post toward you. The line on this face will be cut partway, so use that cut to guide the sawblade as you cut the next face. Roll the post one more time to make the final cut. The last side is the toughest because you need to make sure the sawblade doesn't bind as you cut through. It's a good idea to have another person catch the weight of the offcut if it's long. If the offcut is short, you can just let it drop to the ground.

To cut a 6×6, mark the length and draw lines on three sides. Roll the beam until the first of the three lines is facing up. Cut that line, then roll the beam to the next face and cut that line. The third pass with the saw completes the cut.

Set the bottom of the post in its bracket and push it into position. Then check the post for plumb.

When the post is plumb, toenail through the LVLs and into the post on both sides of the corner using two or three 16d nails.

beneath the LVL. If the fit is a little tight, keep knocking the prop toward vertical until the post slides in. When the post is in place, tap the bottom of the prop out to lower the LVL beam. Now plumb the post. When the post is plumbed and set, toenail through the LVL and into the top of the post on two sides. Move the prop and repeat the process for the rest of the posts.

Install the rafters

The ridge on the other roofs of the house served as the place where pairs of opposing rafters could lean or push against each other for support and to create the roof planes. But the porch roof is a little different because the rafters have no mates to push against. Instead, the tops of the rafters are carried by the ledgers, so they have to be supported by metal hangers. Unlike standard hangers, these are adjustable with a bottom section that hinges to match the angle of the rafter. The tops of the rafters won't be easily accessible after the roof is sheathed, so after laying out the rafter positions on the ledger, nail the hangers on the layout lines. The rafter layout is taken from the joist layout below, but because the rafters are nailed to

After attaching special metal rafter hangers to the ledger, nail in the rafter for the porch gable.

Cut the porch gable sheathing to fit over the rafter as well as the beam below and slide it into place.

the sides of the ceiling joists, the rafter layout falls on the other side of the layout lines.

Cut the porch rafters as you did the rafters for the rest of the house, using the pattern to mass-produce them. The first rafter to go in is for the gable end. When that rafter is installed, cut sheathing for the porch gable and drop it into place. The sheathing extends down to cover the LVL beam at the end of the porch.

Now you can install the rest of the rafters. They're short, so the installation is a one-person job. Like the rafters in the other roofs, snug the bird's-mouth cuts to the side of the LVL, and then toenail the rafters to the LVL beam with two 12d nails. Drive a couple of 12d nails into the ceiling joist as well.

When all the bottoms of the rafters are nailed off, go back, bend the bottom flange of each metal hanger up to its rafter, and nail the hangers to the rafter.

Add triangular nailers above each rafter to support the sheathing.

After installing the first course, run the second course up to the main roof sheathing and around the dormer.

Because nailing the hangers requires a different nailer, it makes more sense to go back for this part rather than trying to juggle two nailers and two hoses as you move down the roof.

Sheathe the roof

The design of the house created a small problem that had to be resolved before we could begin sheathing. To make the porch ceiling the proper height with a 6:12 pitch roof, the top of the ledger holding the rafter tails sat almost 3 in. above the main roof plane. To blend the porch roof sheathing into the main roof sheathing, we made wedge-shaped nailers and installed them above each rafter. These little wedges are easy to figure.

Set a straightedge on top of the porch rafters and slide it up until it meets the main roof sheathing. Mark the end and then measure to the back of the ledger. Figure the height you need at the ledger and then make your wedges (see "Cutting Small Pieces" below).

The porch roof sheathing went on much like the sheathing for the rest of the house. After snapping a guide line, we started at one end with a half sheet left over from earlier sheathing. Partially drive nails into the ends of the rafters to hold the sheets in place until they can be nailed in, but be sure never to stand on a sheet unless it is nailed securely. As you continue the sheathing, remember to stagger the seams. Cut an L-shaped piece of sheathing to wrap around the corner of the dormer. It's never a good idea to align a sheathing seam with the wall of a dormer. Not only does that create a weak point but it also increases the risk of water leakage.

Cutting Small Pieces

Once you have the dimensions of your wedges, it's easy to mass-produce them in pairs. Clamp or nail one end of a 2× to sawhorses. The height of the wedge is the width of the stock you need. So if the height is 3¼ in., rip the 2× to that width. Now mark the length of the wedge with a square line across the 2×. Connect the opposite corners with a diagonal line and rip down the center of that line to account for the width of the saw kerf. One wedge is cut as the diagonal rip intersects with the stock-width rip. Cut across the square line for a second wedge. Repeat that process for as many wedges as you need. By the way, if you use a wider 2×, you can make more than one stock rip and cut multiple pairs at one time.

We had to fit a dummy rafter tail into place to provide a trim nailer at the end of the house. Pieces of water-proof membrane slip in to protect the intersection of the porch roof and the main roof.

One item to complete the porch framing actually had to do with the main roof. The shortened rafters were installed before the exact location of the porch was determined, and the porch roof ends 2 ft. before the end of the house. That meant that one rafter in the main roof didn't have a tail. No problem: We just nailed a dummy rafter tail into place. The tail provides a necessary nailer for the trim details. Where the porch roof intersects the main roof, lay in some waterproof membrane to seal the intersection.

Button up the porch

There were just a couple of items left to finish the porch and bring it up to speed with the rest of the house. The first thing was to install the rest of the porch floor joists. Cut them to length and tack them into place, then slip on the joist hangers to complete the installation. When the floor joists are in, lay plywood sheathing on top of them to serve as a temporary platform. Then install the joist hangers on the porch ceiling joists. The finish porch ceiling and the

finish porch floor should be installed when the house is nearly complete so they won't be damaged as other work progresses.

As elsewhere, trim should be installed before the roofing. We began by applying felt paper to the porch gables and completing any house trim that ended at the porch. The porch soffit and fascia details matched the rest of the house, but the rake and gable return details were a little different. Instead of the extended rakes with pork chop returns, the general contractor decided to let the fascia return in a horizontal band across the porch gables. The upper end of each rake ends in a long point where the porch roof blends into the main roof. Before we installed the trim, we had to protect the intersection with lead flashing and water-proofing membrane. And the horizontal band created a little more work for the siding crew because they had to add drip flashing to protect it before they sided the gables.

Before you declare that the house shell is complete, go around the house and check for any missing details such as unnailed trim or missing joist hangers. When you're satisfied that all is as it should be, take a breather—you earned it.

There are various ways this porch could have been trimmed out. The general contractor opted for fascia returned to the house in a horizontal band, with rake trim ending at the house roof.

AFTERWORD

The book is about building the shell of a house, but there's a lot of work that happens after the shell is complete. In many cases, the framing crew will turn the work over to other contractors, but in this case they stuck around to install the siding and roofing. The skylights went in as the roofing was installed, and the windows and doors went in with the siding. They also took care of details such as installing a steel bulkhead door on its foundation for direct access to the basement. By the time winter closed in on New England, the house was tight to the weather. Inside, there were a few framing odds and ends to take care of. One of Murphy's Laws of Framing dictates that in every house, at least one tub drain will land directly over a floor joist. Murphy was no stranger at this project, and the crew had to undo some earlier work to gain clearance for the second-floor tub drain. After removing a section of the offending joist, they installed headers and joist hangers to support the cut ends. This sort of work is inevitable when building a house, so you might as well anticipate it.

When the crew finally packed up and moved to their next project, the systems contractors moved in: electricians, plumbers, and HVAC contractors. Without heat in the house, winter makes it tough to finish out the interior unless you *enjoy* working inside a refrigerator. Once the systems were finished, the insulation contractors rolled in. The drywallers arrived next, but before the framing was covered up for good, general contractor Joe Iafrate went over every square inch to make sure nothing was amiss. Drywall eventually gave way to finish carpentry, painting, flooring, and cabinets. Slowly but surely what began as stacks of lumber became a warm and inviting home.

INDEX